THE Home Owner HANDBOOK OF
CARPENTRY
and WOODWORKING

By ROBERT BRIGHTMAN Illustrated by HENRY CLARK

EISINGER PUBLICATION 101

PUBLISHED BY
EISINGER COMMUNICATIONS, INC.
42 CARLTON PLACE, STATEN ISLAND, N.Y. 10304

LAWRENCE A. EISINGER: *President/Editor-in-Chief*
LAURENCE EISINGER, JR.: *Editorial/Marketing Associate*

RAY GILL: *Editor*	•	**JACQUELINE BARNES:** *Associate Editor*
SILVIO LEMBO: *Art Director*	•	**JOHN CERVASIO** *: Art Editors*
HOWARD KATZ: *Production*	•	**ALEX SANTIAGO**

Special photography by Kenneth Brightman
Cover photos by John Capotosto

ACKNOWLEDGMENTS

We gratefully acknowledge the help of the following firms:

- American Plywood Association
- Andersen Company
- Armstrong Company
- Aromatic Red Cedar Association
- Black and Decker
- Boise Cascade
- Carrier Corporation
- Douglas Fir Plywood Association
- Dow Corning
- Dremel Tools
- Dupont Corporation
- Edmund Scientific
- Faultless Caster Company
- Formica Corporation
- Georgia Pacific
- Johnson's Wax
- Lennox Corporation
- Marlite Company
- Masonite Corporation
- McCulloch Company
- Minwax
- Montgomery Ward
- National Oak Flooring Association
- National Paint and Varnish Association
- Olin Company
- Ponderosa Pine Association
- Rockwell Company
- Rohm & Hass
- Savogran Company
- Sears Roebuck
- Sherwin Williams Company
- Stanley Works
- 3M Company
- Toolco, Inc.
- Vise-Grip
- U. S. Plywood
- Western Wood Molding and Millwork Association
- Western Wood Products Association

ABOUT THE ILLUSTRATOR

Henry Clark has been illustrating how-to-articles for 25 years. He is not only a professional artist, but a knowledgeable do-it-yourselfer. He *knows* how to use all tools, he *knows* how to work with all building materials because he has been building homes inside and out for these past 25 years.

Unlike most publishing procedures, when Henry started illustrating this book we merely gave him chapter-by-chapter headings and allowed him to use his fertile imagination. The beautiful, fact-filled, informative illustrations that are featured are the result. Thank you, Hank, for making this a great book!

Larry Eisinger, Editor-in-Chief

Dedicated to my wife Mollie, who sutured all split infinitives and snipped off many dangling participles.

Robert Brightman

No part of this publication may be reproduced without written permission from Eisinger Publications, Inc., except by a magazine, newspaper, newsletter, TV or radio station who wishes to refer to specific chapters while preparing a review.

Copyright MCMLXXIV, MCMLXXIX Eisinger Communications, Inc.
All rights reserved. Printed in U.S.A.

CONTENTS

CHAPTER		PAGE
	HAND TOOLS	
1	Hammers	6
2	Saws	8
3	Utility Saws	10
4	Screwdrivers	12
5	Planes	13
6	Squares	14
7	Measuring Devices	15
8	Levels	16
9	Hand Drills	17
10	Wood Chisels	18
11	Miter Boxes	18
12	Awls	20
13	Clamps and Vises	20
14	Miscellaneous Hand Tools	22
	PORTABLE POWER TOOLS	
15	The Drill	26
16	The Circular Saw	30
17	Sanders	32
18	The Saber Saw	34
19	Utility Saws	36
20	The Plane	37
21	The Router	38
	STATIONARY POWER TOOLS	
22	The Circular Saw	40
23	The Radial Arm Saw	42
24	The Band Saw	44
25	The Drill Press	45
26	The Sander	46
27	The Jig Saw	47
28	The Shaper	48
29	The Jointer	49
30	The Lathe	50
31	Lumber and Plywood	52
32	Wood Paneling	56
33	Wallboard, Hardboard and Plastic Laminates	60
34	Planning Your Workshop	64
35	Building a Workbench	68
36	Splices and Joints	70
37	House Framing	76
38	How to Hang Doors	84
39	Installing Kitchen Cabinets	88
40	How to Cover Countertops	92
41	Facts About Fasteners	94
42	Sanding Techniques	98
43	High Points on Ceilings	100
44	Build This Cedar Closet	104
45	Advice on Adhesives	106
46	Keeping Tools Sharp	108
47	Eight Ways to Hang Shelves	112
48	It's the Finish That Counts	114
49	Paints and Enamels	119
	Index	124

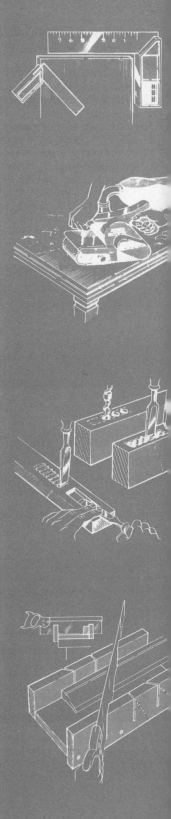

TWELVE ESSENTIAL HAND TOOLS

Hammers

1. Hammers. The basic tool for the carpenter, as well as the homeowner, is the hammer. The claw hammer is used for nailing, nail pulling and dismantling work. Its face is usually bell-faced (convex) to minimize marring when a nail is driven flush. A good hammer will have a tempered rim to reduce the chances of chipping. Equally useful is the ripping hammer. This has a comparatively straight claw and is very handy for removing floor boards and similar work where a prying action is required. Avoid cheap hammers made of cast iron; they are bound to shatter under severe work.

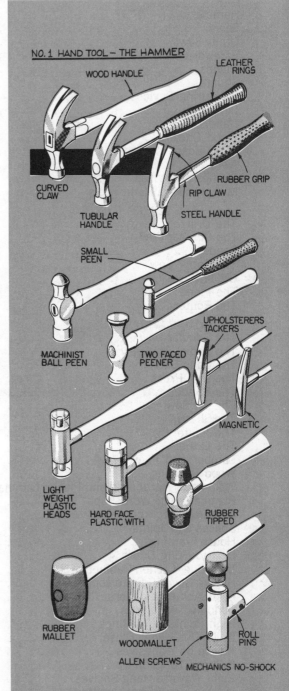

NO. 1 HAND TOOL – THE HAMMER
- WOOD HANDLE
- LEATHER RINGS
- CURVED CLAW
- TUBULAR HANDLE
- RIP CLAW
- RUBBER GRIP
- STEEL HANDLE
- SMALL PEEN
- MACHINIST BALL PEEN
- TWO FACED PEENER
- UPHOLSTERERS TACKERS
- MAGNETIC
- LIGHT WEIGHT PLASTIC HEADS
- HARD FACE PLASTIC WITH
- RUBBER TIPPED
- RUBBER MALLET
- WOODMALLET
- ALLEN SCREWS
- ROLL PINS
- MECHANICS NO-SHOCK

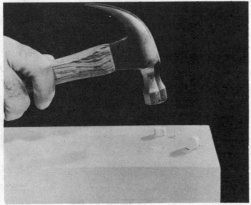

Hammer shown above is a claw hammer, the No. 1 hand tool. They come in various weights and with wood, steel, and fiber-glass reinforced handles. Ripping hammer has straight claw, good for prying.

Another useful hammer is the short-handled sledge. Use it for driving case-hardened masonry nails and where an extra heavy hammer is required. Note the use of goggles, a must when doing this work.

Twelve essential hand tools can take care of most of your woodworking tasks. Use this comprehensive guide on how to choose and use them.

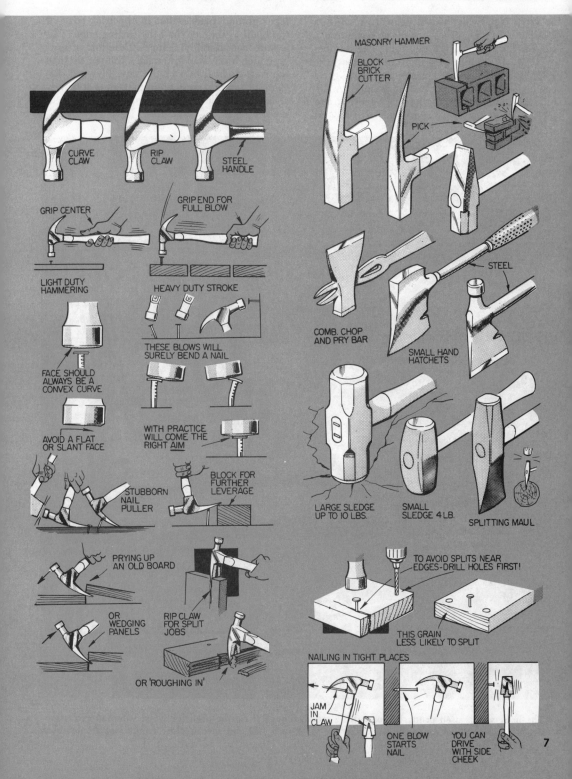

| TWELVE ESSENTIAL HAND TOOLS | # Saws

2. Saws. You will need at least two saws—a crosscut saw for cutting across the grain and a rip saw for cutting with the grain. The number of teeth or points per inch determines the smoothness of the cut. For average work, get a crosscut saw with eight teeth per inch. A 26-inch rip saw will generally have 5½ teeth to the inch. A good saw (rip or crosscut) will have a taper ground blade—that is, the back edge of the blade is thinner than the cutting edge. This feature considerably reduces the effort of sawing. Such a saw will also have a thinner tip than the butt or heel end of the saw. Plywood, because of the crisscross grain pattern of its several layers, should always be cut with a crosscut saw.

The hack saw is used primarily for cutting metal, though it can be used for wood when an extra fine cut is desired. When sawing metal, hold the work in a vise and saw as close to the vise as possible.

The No. 2 hand tool, after the hammer, is the saw. Your first choice should be a crosscut saw; the next a ripsaw.

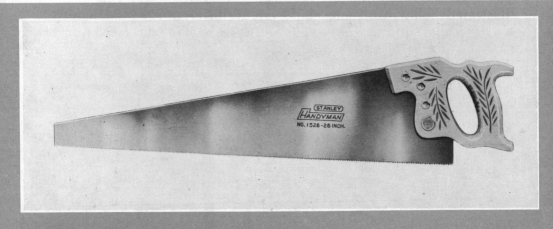

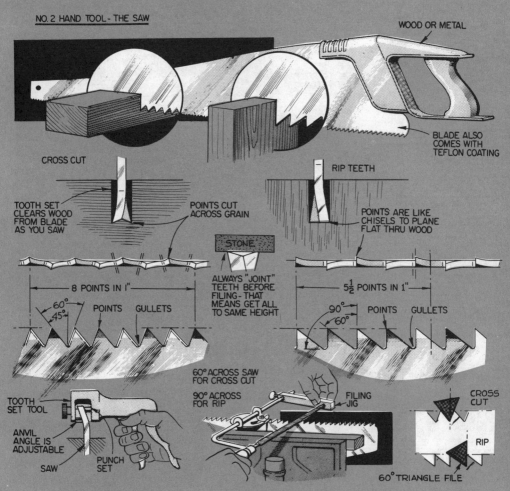

The Utility Saws

Illustrated below are some of the many different kinds of hand saws you may find useful in your work. How many do you have?

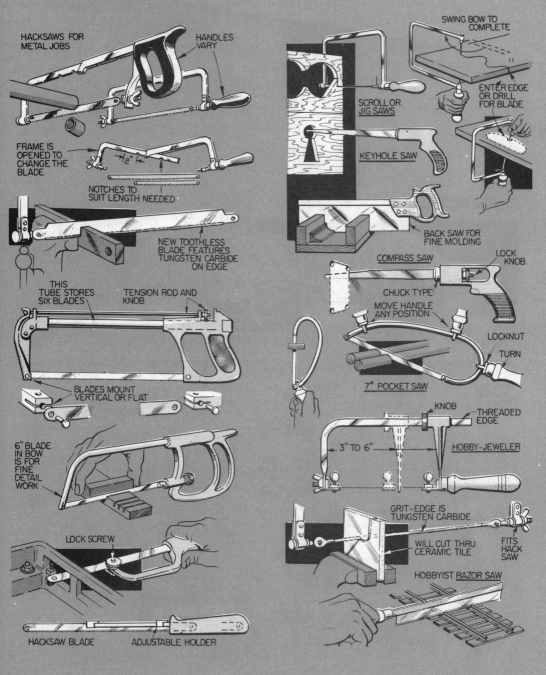

THE COPING SAW

The coping saw has an extra-wide throat and is used for cutting curved, deep openings as shown in photo. The blade can be mounted in the frame so that cutting can be done on the pull stroke or the push stroke, whichever is more convenient for the operator.

THE BACK SAW

The back saw is primarily used for fine cabinet work where an extra-fine cut is necessary. Here the operator is using a number of clamps to hold the work steady and a block of wood as a guide in order to make a dado cut to an exact depth and full width.

MITERING

It is the steel backbone of the back saw that contributes to its precision when used by a careful worker. Back saws are available in various lengths, the longest, 24-in. is generally used by professionals in a steel miter box. A back saw, about 15-in. long is best for the home handyman.

| TWELVE ESSENTIAL HAND TOOLS | A selection of screwdrivers will make work easier, prevent damage. |

Screwdrivers

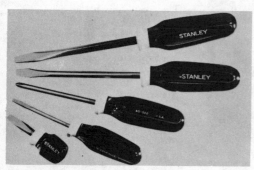

A good investment is a set of screwdrivers which should include a Phillips as well as a stubby model. Tools are a lifetime investment—even if your neighbor does not return them—buy the best.

3. **Screwdrivers.** Get a set of at least four. It is pointless owning only a single screwdriver—even if it is a good one—as screws come in so many different sizes. The wrong screwdriver will either chew up the slot or else slip out and damage the work. You should have at least two Phillips-type screwdrivers, a standard blade tip for general work, and a long slim screwdriver sometimes called a cabinet screwdriver. Also handy to have are offset screwdrivers, stubby models, screw-holding screwdrivers, a jeweler's screwdriver for fine work, and last, but not least, the Yankee spiral ratchet screwdriver which drives—and removes—screws by means of a pushing action on the handle.

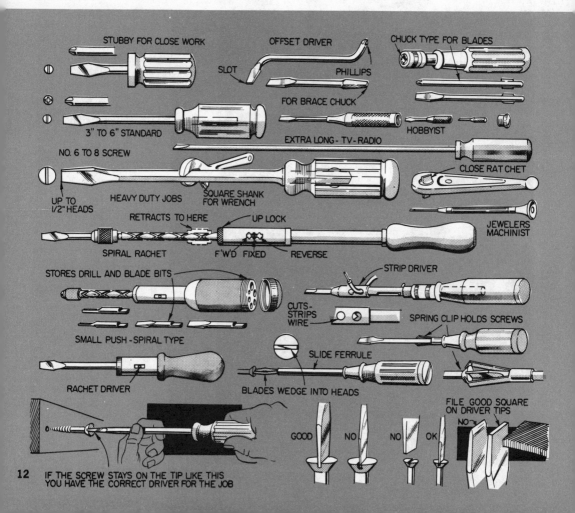

12 IF THE SCREW STAYS ON THE TIP LIKE THIS YOU HAVE THE CORRECT DRIVER FOR THE JOB

Planes

Planes are the shapers and the trimmers of your tool collection.

4. Planes. Planes are used to trim wood to an exact size, to bevel edges, to smooth out irregularities, and even to make moldings. The block plane is about six inches long; it is always held in one hand (the other hand is used to steady the work, if necessary) and its blade is mounted at a shallow angle so that it can cut across the grain without splitting the wood. Next in size comes the bench plane, about ten inches long (usually the first choice of the do-it-yourselfer); then the jack plane, about 14 inches in length; and finally the jointer plane, 18 to 22 inches in length. In addition there are many specialty planes used to cut rabbets and grooves in boards. There is even a tiny plane suitable for the model-maker—it is called a trimming plane. It has a narrow blade, an inch in width.

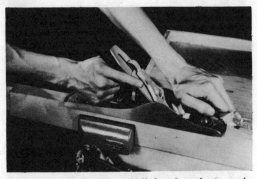

The most fascinating of all hand tools is probably the plane—chiefly because it is one of the few hand tools that has so many internal adjustments. Always make sure work is securely held.

The Surform (pronounced Sir Form) is a sort of poor relation to the plane, but it has many advantages over the plane. It requires no sharpening or adjustment and comes in various styles.

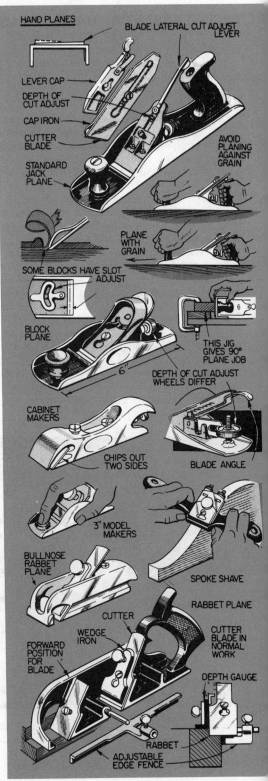

TWELVE ESSENTIAL HAND TOOLS

Squares

5. Squares. Squares come in many sizes and types. The most common, and the most useful for the handyman, is the combination square. It has an overall length of 12 inches, can also be used as a ruler and a level, and has a 45° setting, a 90° setting, and a built-in scratch awl. The blade is grooved so that it can be set and locked at any desired length to measure and indicate depth. The try square is used to determine exact right angles and the "levelness" of work. The miter square is similar to the try square, but its handle at the point where it meets the blade is cut at a 45° angle for measuring, marking and testing 45° miter joints. One of the most important squares for the carpenter and builder is the rafter or roofing square. This is a steel square with one leg 24 inches long and the other leg (the tongue) 16 inches long. It is an extremely useful tool for marking rafter cuts according to roof pitch, determining brace length between two points, finding the center of a circle, laying off angles, and similar problems in building construction.

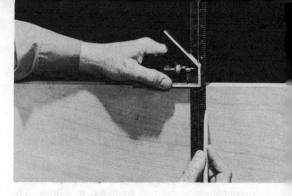

The combination square is chiefly used for what its name implies—marking a line that is absolutely square to the handle of the tool. Has a built-in level and is also used for marking 45° miters.

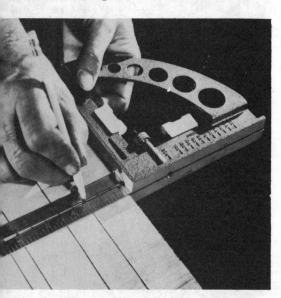

This combination tool can be used for marking, as a compass, level, nail and screw gauge, depth gauge, as a square, and for measuring angles. The manufacturer maintains there are 11 uses for it.

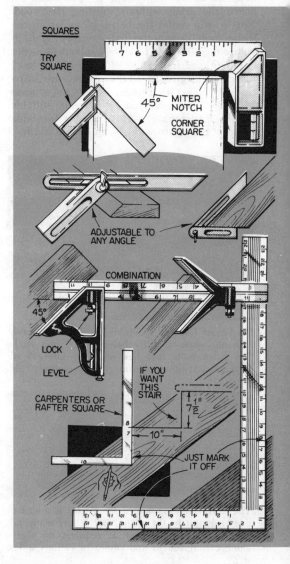

Measuring Devices

6. Measuring Devices. A good, all-around measuring device is the flexible steel tape. Get one at least ten feet long and make sure the right-angle clip at the end is self-adjusting for inside as well as outside measurements. Folding wood rules extend to six or eight feet. Some have a sliding brass extension for determining inside measurements; some are marked off at 16-inch intervals to expedite stud placement. The advantage of the folding zig-zag ruler is that it is stiff enough to measure across horizontal openings without collapsing.

The most popular measuring device is the steel tape. Pocket models come in lengths up to 12-ft., widths from ¼-in. to ¾-in. Most are spring-loaded so that blade automatically retracts when the side of the case is pressed to release the holding device.

Second in popularity is the folding zig-zag ruler. Made of wood, it can span a 6-ft. opening, supported only at one end. Some are made with special markings to indicate correct spacing for wall studs.

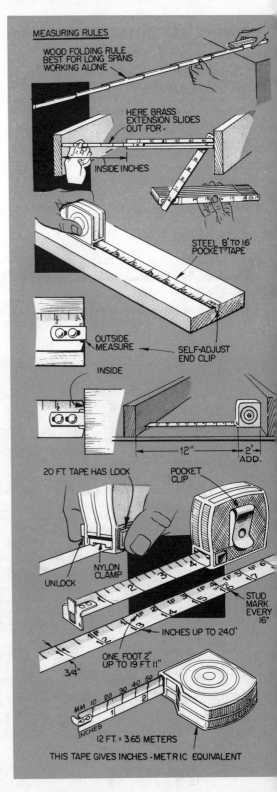

| TWELVE ESSENTIAL HAND TOOLS | Levels |

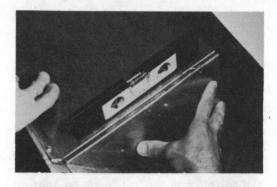

7. Levels come in at least a dozen different types and sizes. A good level for the do-it-yourselfer is the 24-inch carpenter's level. It has at least two vials, one to determine horizontal "levelness" and the other for determining the "uprightness" (plumb) of a vertical member. Other levels useful for the carpenter and handyman are the torpedo level, about nine inches long; the line level (suspended on a string to span a long distance); and the pocket level, usually hexagonal so it will not roll; it is about four inches long.

Some representative levels are shown above. Levels are made of wood, aluminum, and steel. There are levels for "leveling" masonry, lathes, and pipes; some even have adjustable vials for use in ship-building.

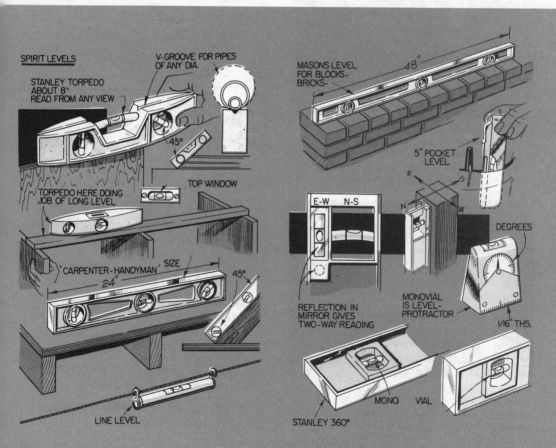

Hand Drills

8. Hand Drills and Braces. The electric drill has to a large extent replaced the hand drill and the carpenter's brace. However, the latter two tools are still useful. Their chief advantage, of course, is their ability to be used without the need for an electric cord. A plus for the brace is the tremendous torque which it can generate because of its wide, 10- to 12-inch sweep. A comparatively little known drilling device is the Yankee push drill. It is much favored by professional carpenters and cabinet makers. It is used to drill small holes, not larger than 11/64-inch. A hollow handle stores the bits. Pushing on the handle drives the bit into the work; a spring forces the handle back for the next stroke.

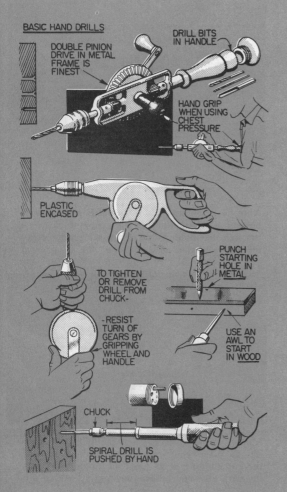

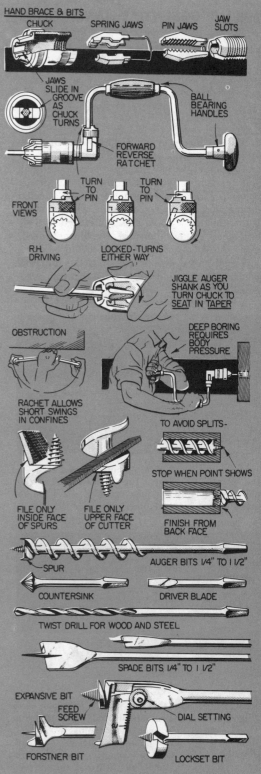

TWELVE ESSENTIAL HAND TOOLS

Wood Chisels

9. Wood Chisels. A set of chisels is a must—no sense buying them individually. A ¼-, ½-, ¾- and 1-inch chisel will take care of practically all your carpentry work. Get them in a plastic wrap-up case so their edges will be protected during storage. Do not confuse "wood" chisels with cold chisels. Cold chisels are used for cutting metal, though there are also all-steel chisels made for woodworking. Chisels should be sharp and well-honed before use. It is the dull tool which requires unnecessary force that causes accidents.

Using a wood chisel to clean out the wood between two parallel saw cuts. The beveled side of the chisel should always face up. Use one hand to guide the chisel and other hand for driving it.

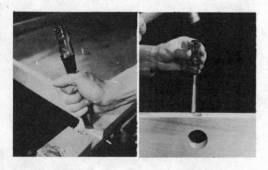

Above, left: how the side of the chisel is used to clean out the shoulders of a dado cut. Right: a chisel is used to deepen the opening for a door lock. This operation is called mortising and can also be done with an electric router. Score deeply the areas to be cut to avoid splintering.

Miter Boxes

10. Miter Box. The miter box—and its attendant back saw—are absolutely indispensable for cutting 45° miters in molding. A sturdy maple miter box and back saw will cost you less than ten dollars. In addition to mitering, the miter box is also used for making an exact right angle cut across a piece of molding or wood, narrow enough to fit into the box.

Note how a scrap piece of wood is placed at the bottom of the miter box to assure a clean cut through the molding and at the same time it protects the bottom of the miter box from wear.

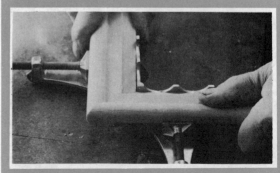

A miter clamp is an extremely handy gadget if you like to make your own picture frames. It holds the cut molding at an exact 45° angle for fastening with glue and brads. Unclamp after glue sets.

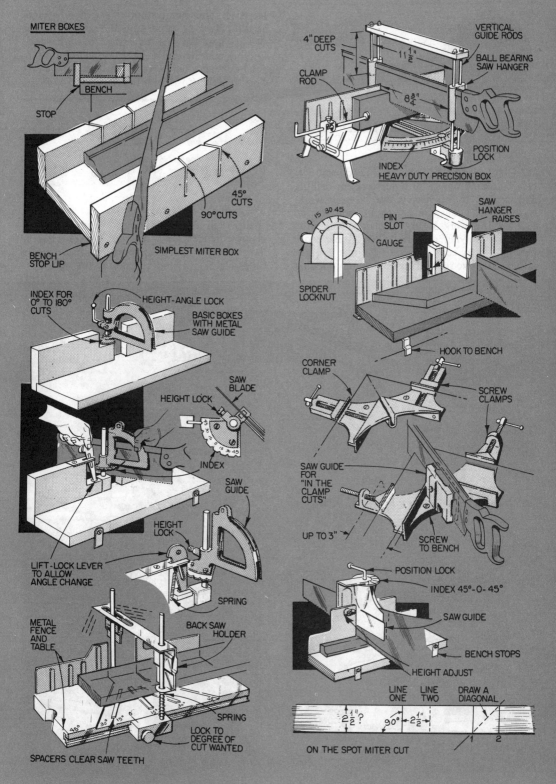

Awl

11. Awl. This handy and inexpensive tool is used for marking, scratching and scoring. It can also be used for starting screws in soft wood. Get one with flattened sides so it will not roll. A good one has a tempered steel shaft and a hardwood handle. Cost? Less than a dollar.

TWELVE ESSENTIAL HAND TOOLS

Clamps and Vises

12. Clamps and Vises. You will need some sort of a clamping device to hold work for drilling and sawing. If you have a vise on your workbench, fine. Otherwise a C-clamp can be used to hold the work steady. C-clamps are often used for applying pressure when gluing work together. They come in sizes from one to eight inches. The size refers to the maximum opening —not the depth of the throat. For extra long spans, such as when gluing furniture, use pipe clamps. These clamps fit over standard ½- or ¾-inch pipe (one end must be threaded) to form a gluing clamp as long as the pipe.

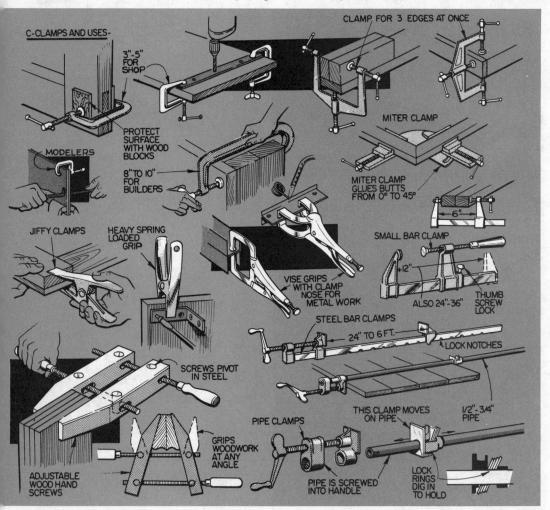

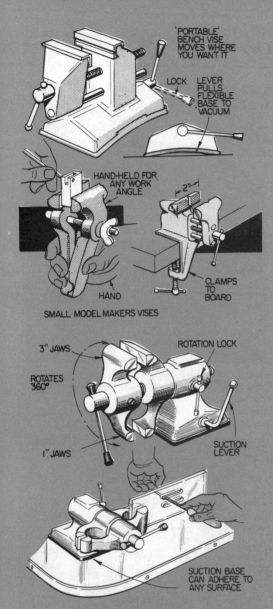
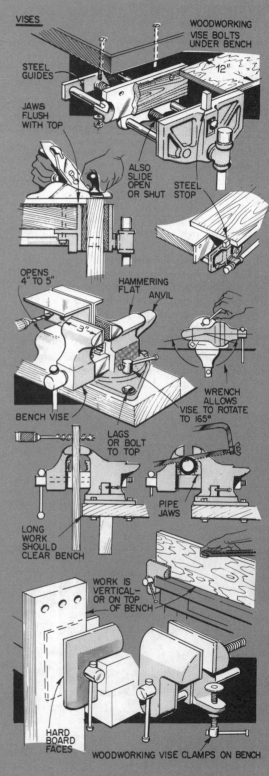

THE VISE

The vise is the third hand in the workshop. After making a workbench (see page 68) the next step is to fit it with a vise. You can either use a woodworker's vise or a machinist's vise—or both, one at each end. If you opt for only one vise, get the machinist's vise. It can be used for holding wood safely if you pad the jaws with hardboard or wood. The woodworker's vise is mounted below the workbench, while the metalworking or machinist's vise is mounted on top of the workbench.

Miscellaneous Hand Tools

In addition to the tools we've listed, there are a dozen or so small miscellaneous tools that are extremely handy—and inexpensive.

- A pry bar or ripping bar for removing molding, for installing and removing floor boards and for general dismantling work.
- A mason's line and chalk for snapping long, straight lines.
- A nail set for driving finishing nails and brads below the surface of the work.
- A pincer for removing nails.
- A wing divider or a compass for scribing circles and arcs, and setting off equal measurements.
- A utility knife for general cutting and scribing.
- A putty knife for what the name implies.
- A wall scraper or joint knife. This is a wide-blade version of the putty knife, used for wallboard work.
- A plumb bob for determining vertical lines.
- Pliers and wrenches, while not true carpenter's tools, are essential for a good deal of the work around the house. Get pliers

Handy all-purpose plier will cut electric wire and strip it. It will also cut five sizes of bolts.

Handy wide-blade knife is used to remove labels and paint from window glass; its handle is hollow for storage of extra blades and of course it can be used for opening cartons and general cutting.

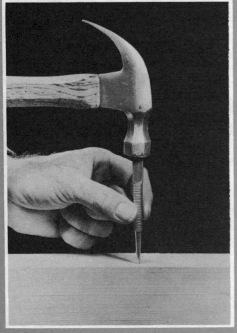

The nail set is used to drive a finishing nail, or a brad slightly below the surface of the work. Then a filler is used to cover the nail head, thus making it invisible, and ready for staining and finishing. Caution: when the nail set is nearing the surface of the work, hold it steady so it won't slip off the head.

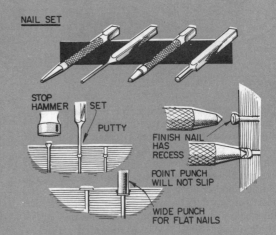

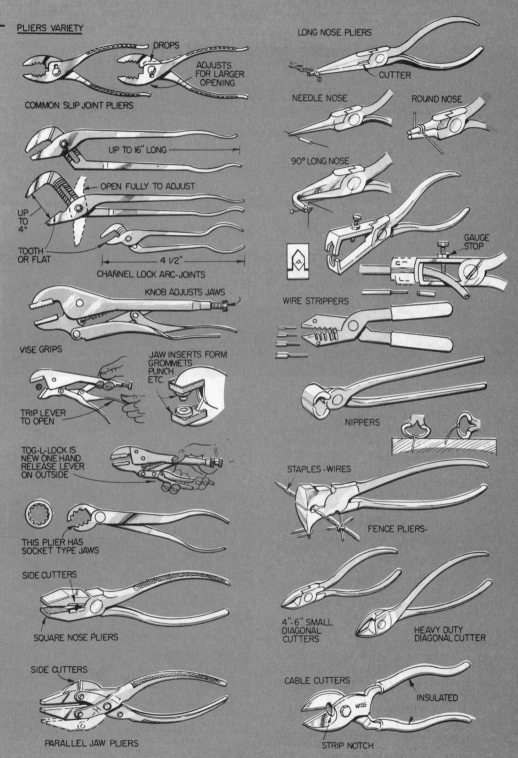

MISCELLANEOUS HAND TOOLS

The Surform, kissing cousin to the plane, being used to trim edges of a child's chair seat to fit within the frame. While the Surform's blade can not be sharpened, replacement blades are available.

that can also cut heavy wire. As for wrenches, a pipe wrench (Stillson) and an adjustable open end wrench will do for starters. A combination open-end box wrench set is very handy.
- Locking pliers, the type sold under the trade name of Vise-Grip, are very handy tools.
- A marking gauge, used for scoring perfect parallel lines along the edge of a board.
- A hacksaw, a coping saw and a compass saw, for those odd cutting jobs the crosscut and rip saws cannot handle.
- A countersink for use with a hand drill, electric drill or brace, so that flathead screws can be driven flush with the work surface.
- A Surform tool. This handy tool, made by Stanley, can do most of the work the more expensive plane does. Its advantage is that it less apt to spoil work in the hands of an inexperienced user. It is made in several types and sizes.
- A file and a whetstone for keeping your tools sharp.
- A rasp and some sandpaper in various grades.

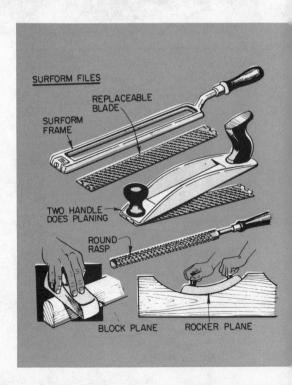

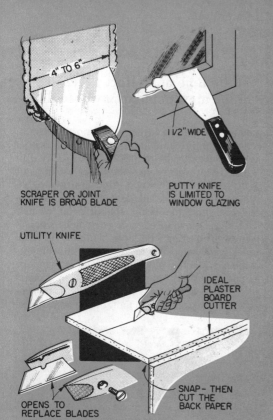
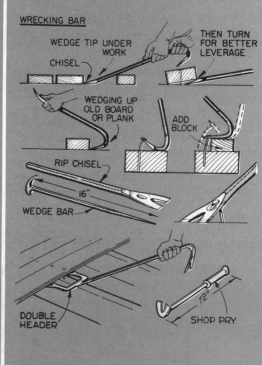
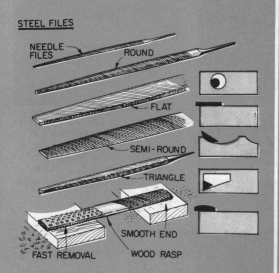
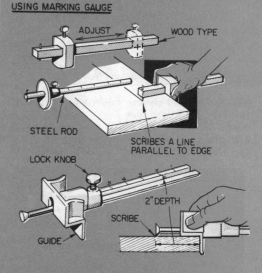

PORTABLE POWER TOOLS

The Drill

These tools are designed to cut down on the time and effort necessary to do a job. Following are descriptions of the most popular power tools.

If you have planned any fair-sized projects, such as the workbench (on pg. 68), you undoubtedly have come to the conclusion that power tools can be of great help in your work. Cutting the 2x4s to length, drilling holes for the bolts, cutting the plywood panels—all are much easier with power tools.

The first power tool you should own is the **quarter-inch drill**. The size designation refers to the drill's chuck capacity. Power drills also come in 3/8-, ½- and even ¾-inch size (the last used primarily for industrial work). If you intend to take your do-it-yourself work seriously, you might want to invest a few more dollars in your initial purchase and get a 3/8-inch drill, a good choice for all-around work. A quality drill will have a reverse gear—useful for backing out a drill bit in tough wood—and also a variable speed feature. The harder you press on the trigger-switch the faster the drill bit rotates. The initial slow speed is a great help when it comes to starting the hole without the drill point wandering. There are many attachments available for the electric drill—saws, sanders, grinders, buffers, etc. However, as a general rule, it is best to buy a tool designed for a specific purpose rather than hanging on accessories.

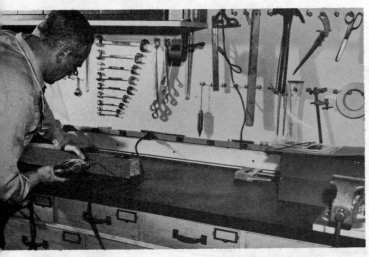

The No. 1 small power tool in your workshop is the ¼-in. drill. However, for a few more bucks, you can buy a 3/8th-in. job, well worth the extra cost. You can buy a fairly respectable drill for $15—and up. The "up" depends upon the bearings, as better drills use many ball or roller bearings instead of bronze bearings.

This ¼-in. electric drill is cordless, can be used anywhere since power is built in. Perfect for on-site work where availability of electric outlets is a problem, or where electricity may be hazard.

Using the ¼-inch drill for countersinking screw heads in the bottom of the cover of a window seat to be used for storing winter clothes.

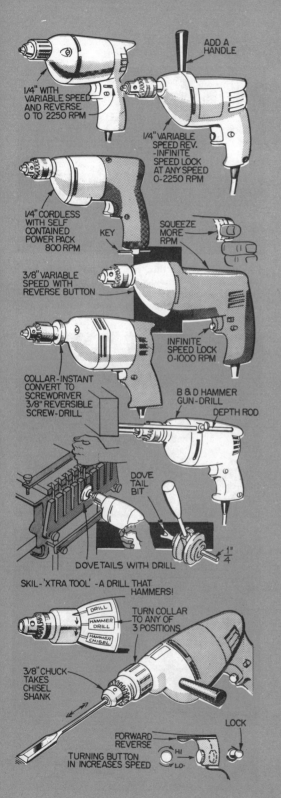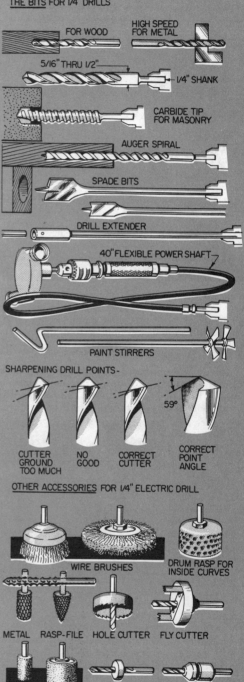

ELECTRIC DRILL ACCESSORIES

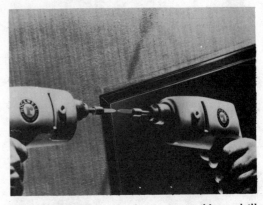

You won't believe this, but it is possible to drill a hole in glass—provided you use a carbide-tipped drill. Use extremely low speed, be careful point does not wander; use water as lubricant.

Some of the many ways the ¼-in. electric drill can be used are illustrated on these pages. The photo above shows an operator using depth gauge on his drill so hole will only go to desired depth.

A wire brush powered by the ubiquitous drill does a dandy job of removing rust and scale from metal. Best to grasp the drill firmly with two hands and wear protective goggles over your eyes.

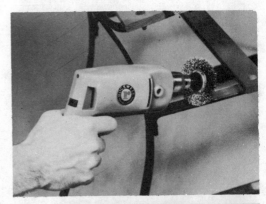

A variable speed drill will enable you to use the drill to drive screws. Pressure on the trigger-switch determines the driving speed. Use slow speed and make sure bit fits screw slot snugly.

Right: use a high speed to drill holes in wood and a slower speed for drilling holes in metal. When drilling holes in metal is called for, use special high-speed drills; they stay sharp longer.

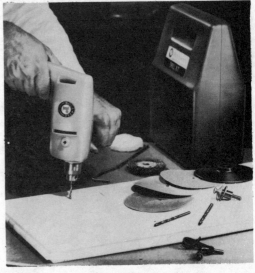

The drill shown above has a special clutch which disengages the power from the screw-driving bit as soon as it meets undue resistance, such as when the screw has been adequately seated.

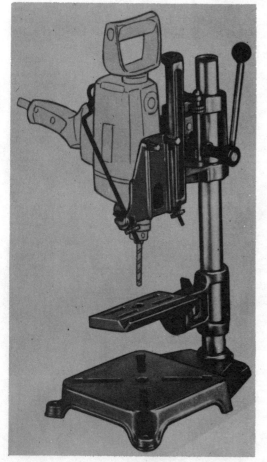

If you own a half-inch drill, you can convert it into a drill press by means of this holding device. Most manufacturers of half-inch drills make such accessories to fit their own drills.

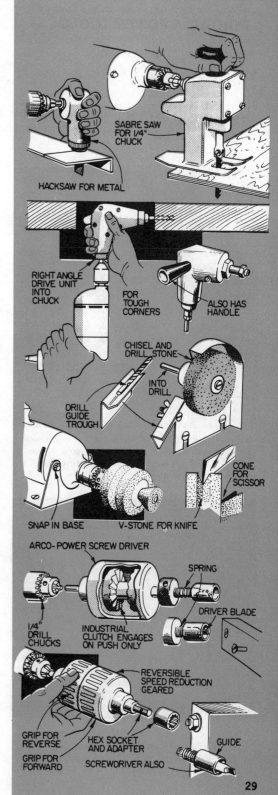

PORTABLE POWER TOOLS

The Circular Saw

This tool revolutionized the home building industry—and it can help you do jobs you never thought possible. Use it carefully.

Crosscutting a piece of wood with the portable power saw. Wood should be supported at each end so it will not bind the saw. Greatest possible depth of cut is determined by diameter of blade.

The next power tool you will probably want to consider—and ultimately buy—is the **circular saw**. With this versatile tool you can crosscut, rip, miter, and make dado and bevel cuts. You can adjust the depth of cut by moving the base up or down as required. You can even make a pocket cut—an opening in the middle of a board—with this saw. The saw you buy will be designated by the size of the blade it uses. They vary from 5½ inches to more than eight inches. A prime consideration is a saw that will cut through a 2x4 at a 45° angle—which means you need a saw with at least a 6½-inch blade.

The prices of circular saws vary from about $25 to around $100. Size of the blade, power of the motor, and the general quality of the saw determine the cost. Get a saw with at least a seven-inch blade and a motor close to one horsepower; controls should be easy to adjust with large-size wing nuts; the guard should have a knob for retracting it for certain operations; front of the saw should have an easy-to-grasp handle or knob to steady the saw as it is pushed through the work.

When a cut is made along length of work, it is called ripping. Always have the good side of the work at the bottom; good side of work is on top when ripping or crosscutting with bench saw.

Before you buy a portable circular saw, heft it to make certain you like the feel of it in your hand —not the salesman's. Controls should be easy to get at and adjust, plus convenient trigger switch.

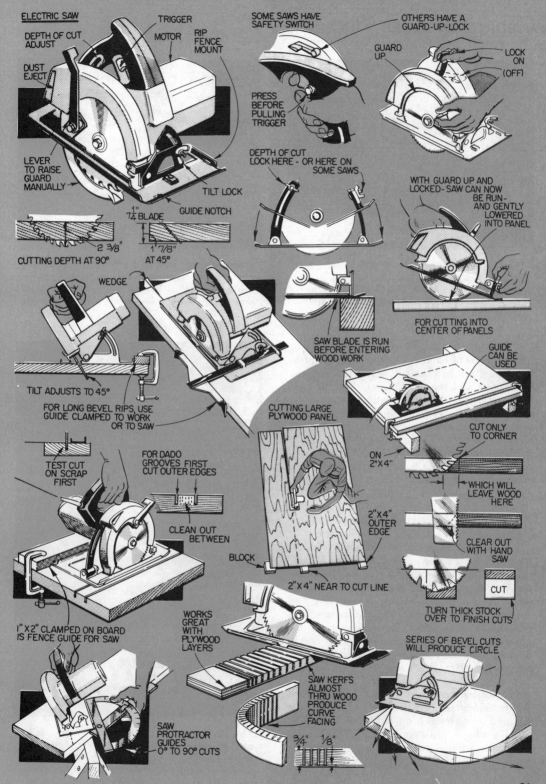

PORTABLE POWER TOOLS

Sanders

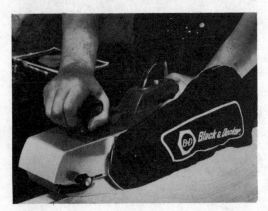

This belt sander is being used with a coarse belt for fast stock removal. Note the vacuum attachment to trap dust. It can also be used with finer-grade belts for smoothing prior to final finish.

Even finishing sanders are now equipped with vac attachments. Best to wear a mask if sander you own does not have a dust-catching device; always sand with grain, do not press into work.

A sander is a good investment. It can save tedious hours of hand work prior to final finishing. There are two main types of sanders: the belt sander and the finishing sander (disc-type sanders that attach to power drills or other power tools are not recommended for finish work). Generally, the belt sander is used for fast removal of stock. A wide variety of belts are available including "extra-coarse," which can be used for removing paint and varnish. Sanders raise a great amount of dust; for this reason many are available with either a built-in dust collecting bag—sort of a miniature vacuum system—or with an attachment so that a household vac can be used as the dust collector.

The **finishing sander** is a different breed altogether. As its name implies, it is used exclusively for finishing work prior to painting, lacquering, or varnishing. A finishing sander can be an orbital type in which the sanding pad moves in a small-diameter circular motion. It is effective when sanding across more than one grain direction such as on miter joints or checkerboard patterns. The other type is the straight-line sander. Here the motion is in a straight line, a tiny movement back and forth. Always sand with the grain when using a straight-line sander. To cover all bets, some sanders can be adjusted from straight-line to orbital sanding by pushing a lever. Most sanders are motor-driven, except for a few economy models that operate by means of a vibrator. This latter type can only be used with alternating current.

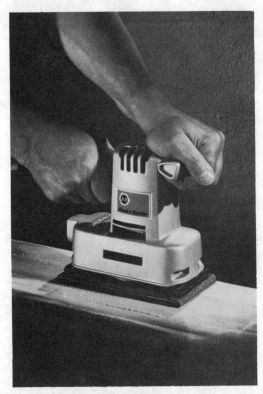

This Black & Decker combination sander can be used for orbital sanding or for straight-line sanding by moving lever in base of sander. Orbital sander is used when work has complex grain.

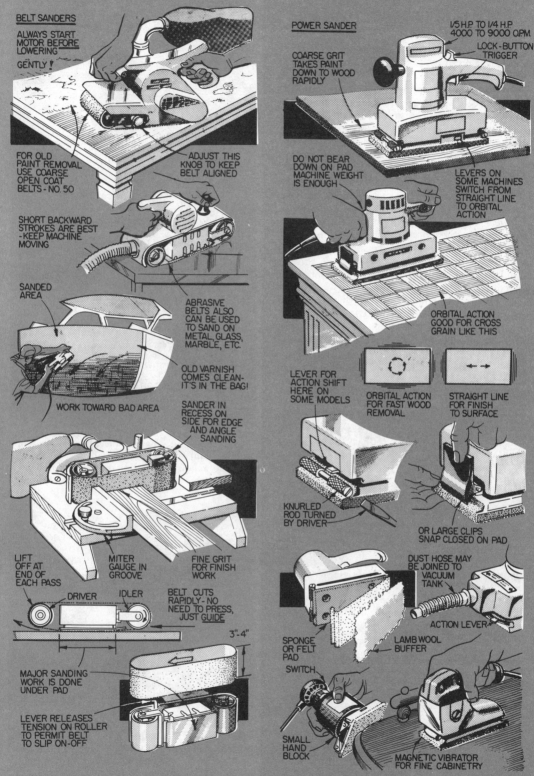

PORTABLE POWER TOOLS — Saber Saws

The **saber saw**, or **jig saw** as it is sometimes called, is used to make straight or curved cuts in wood, hardboard, light metal, laminates and plaster board. A large assortment of blades is available for the saber saw, even knives to cut cloth and paper in bulk. Most saber saws have a tilting base plate for making bevel cuts. A refinement on some of these saws is a blade-turning device handy when working in close quarters. Here is how it works: instead of turning the entire saw to make a tight curve, you merely turn the blade by means of a knob located near the handle. Bear in mind that saber saws cut on the up stroke so always have the "good" side facing down. The slight splintering which sometimes occurs on the top side can be eliminated by applying masking tape along the line of cut.

The up and down motion of the blade in a saber saw is obtained by means of a crank, like a connecting rod, fastened to a gear wheel driven by the motor. There are a few saber saws on the market which operate by means of a vibrator. These should only be used for light work such as model making; they can only be used with alternating current.

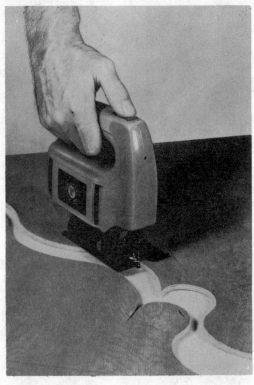

Left. This Sears saw has the ability to cut in all directions because the blade is offset. Its a handy feature for cutting out letters for signs, nameplates and the likes. It also cuts straight.

The side cutter attachment of this saw makes it possible to trim close to a wall without any interference from the housing of the tool. This is a special feature you will find very useful.

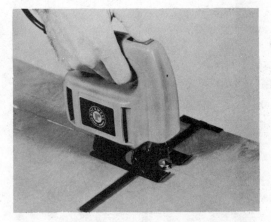

Want to rip a board or panel the easy way? If the distance is less than 8 in., a rip fence will do the same job as a circular saw. You can also cut circles by rigging the arm as shown at right.

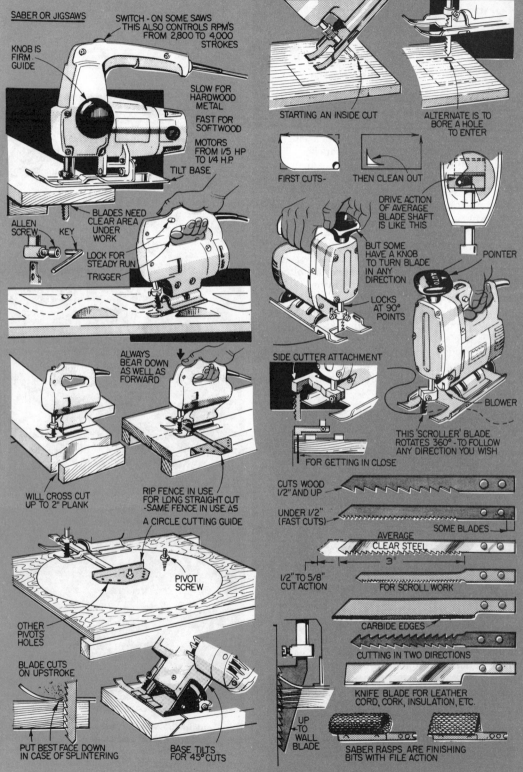

PORTABLE POWER TOOLS

The Utility Saws

The **reciprocating saw** is a sort of larger version of the saber saw. It is designed for heavy duty work, making rough cuts that can be trimmed to exact size later with a circular saw, and for lopping off tree branches. The blades for the reciprocating saw are up to 12 inches and can cut wood up to six inches thick. The blade is mounted parallel to the body of the saw, instead of at right angles as in the saber saw.

While the chain saw cannot really be considered a carpenter's tool, it is extremely handy for removing trees, lopping off branches, cutting down large timbers and general demolition work. The introduction of chain saws weighing less than ten pounds has served to popularize this tool with many homeowners. A chain saw can cut through a 12-inch hardwood log in less than 12 seconds. A saw with a 12- to 16-inch bar is a good choice for most around-the-house work. You can get them powered with a gasoline motor or electrically powered for use with a long extension cord.

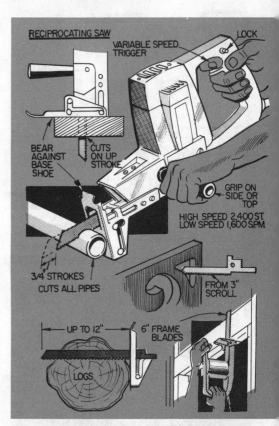

The reciprocating saw is a sort of oversized version of the saber saw. It can accept a blade up to 12 in. long. Good for rough work such as cutting openings in walls and even for slicing logs.

New light-weight chain saws by McCulloch are ideal for making rough cuts in timbers during house construction as their self-contained gas engines permit use before power becomes available.

Here, out in the woods, miles from a power line, an intrepid woodsman and Dave Kirby are building a log cabin with the aid of a light-weight chain saw. Saw is being used to trim half-logs to size.

The Plane

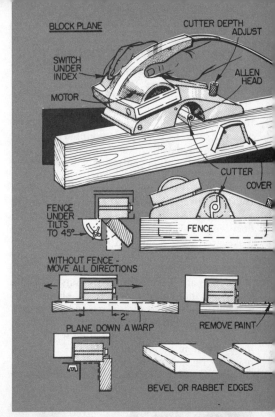

Another portable power tool you may want to consider is the **electric plane**. There are two models available. One is a block plane which, like the conventional block plane, is held in one hand. It can be used for planing the inside of windows and door jambs and for edge planing of sticking doors. Easing sticking drawers is a cinch with this handy tool. A larger version is the electric jack plane. This plane does the work of a hand plane but with much less effort and much greater precision. It planes by means of a revolving cutter, turning at 22,000 rpm. The plane has a fence that projects downward from one side, perpendicular to the bottom surface of the plane. When planing the edge of a door or a board, this fence bears against the side of the work, assuring a perfectly square-edged cut. The fence can be removed for planing wide areas and it can be tilted for bevel planing.

The electric plane is a great time-saver when it comes to the job of fitting a door. Some planes use a router as the power source. The router is fitted into a special holder—and presto—a plane.

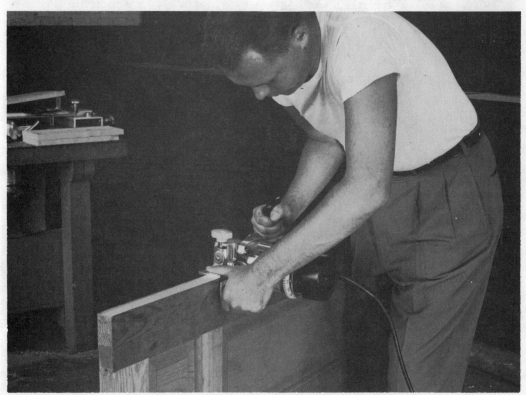

| PORTABLE POWER TOOLS |

The Router

We have left for last one of the most sophisticated and important portable power tools: the **router**. Basically it consists of a high speed motor (about 24,000 rpm) mounted vertically on a horizontal base plate with a chuck at one end to accept various cutter bits. Once the bit has been mounted in the chuck—or collet—it can be moved up or down to cut deep or shallow grooves in the work. It is indispensable for mortising (recessing) hinges in door jambs and door edges. The router can also be used to make grooves and cut dado rabbets. With an attachment it can cut exact circles of practically unlimited diameter.

There are literally dozens of bits made for the router that will enable it to cut shapes suitable for moldings, making V grooves, dovetails, beading, chamfering, ogee curves, concave and square cuts. The drawings illustrate some of the more common router bits and the cuts they make. A router guide is available for cutting a straight line at a set distance from the edge of the work.

How the router is used to cut a decorative edge. The depth of the cut is controlled by raising or lowering the base, which in turn determines the length of the bit; many bit shapes are available.

With a special bit, the router is the best tool for edging plastic laminates. The bit is ground so the top rides the front of the work and does not permit the cutting edge to mar the face.

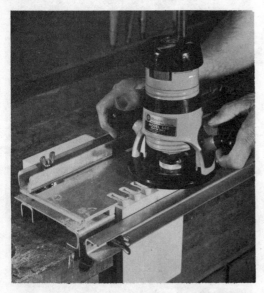

Ever wonder how dovetail joints are made? Only a real craftsman can cut them by hand, but with this special attachment and bit anyone can make dovetail joint for a drawer. Just follow notches.

Rather than chiseling out for each door butt, the router does the job better and faster. A jig is clamped on the door edge and the tool, with a flat bottom bit, is moved back and forth.

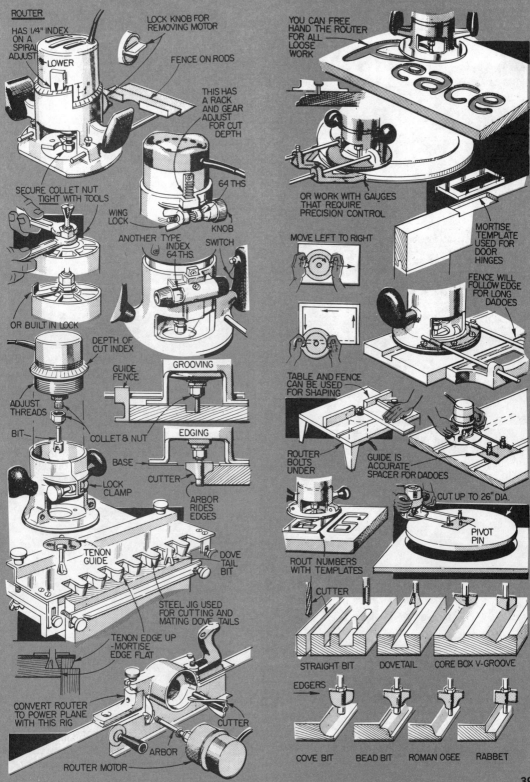

STATIONARY POWER TOOLS

The Bench

The four horsemen of the home workshop are the bench saw (or you may prefer a radial arm saw), the band saw, the sander and the drill press. Other useful stationary power tools are the grinder, the jig saw, the shaper, the jointer and the lathe.

The **bench saw**, sometimes referred to as the table saw or circular saw, is the primary stationary power tool for most homeowners. Its size designation refers to the maximum size blade it will accept. A good all-around choice is an eight-inch saw, but if you can afford it, get a ten-inch model. Power is another criterion; get one with at least a ½-hp motor.

The bench saw can cut lengths of wood to any width; simply set the fence to the desired

Cutting a board along its length is called ripping. The position of the fence in relation to the blade determines the width of the cut. Always stand to one side of the work as shown.

Keep the board pressed against the fence and hook your fingers over the fence, just to make sure they are out of the reach of the blade. Press down firmly on wood and push forward.

The table saw is the workhorse of the shop. Pictured above is an eight-inch saw with an extension on each side to expedite the handling of wide boards, especially sheets of plywood.

The jointer smooths the edges of cut wood so they will make a tight-fitting joint (hence its name). The tool illustrated above is a jointer mounted in combination with a table saw.

Saw

Once you have your own workshop, you'll want to consider power tools. Add them, one at a time, as you need them

distance from the blade. Crosscutting is done with the miter gauge which fits into a groove milled into the table surface. Most saws have grooves on each side of the blade so that you can use the one which is most convenient. Diagonal crosscutting can be done by adjusting the miter gauge. Ripping can also be done at an angle by tilting the blade. The blade can also be lowered or raised to cut any depth for grooving or dadoing.

Make all adjustments with the cord unplugged. Always feed the work into the saw from the front. Start the saw and let it reach full speed before pushing the work into the blade. The blade will slow somewhat during the cutting operation but this is normal. If the blade slows so much that the work starts to smoke, you are pushing the work too fast. Feed the work at a slow, constant rate.

Use the blade guard whenever possible. This should be removed only for those operations that cannot be done with the guard in place, such as dadoing, grooving and edge-cutting of wide stock. Always stand to one side of the blade—never in front of it. And don't wear a tie. These precautions apply to use of most power tools.

For cutting extra-wide grooves, a dado set should be used. This will permit you to cut grooves from ¼-inch to ¾-inch wide by adding or removing chipping blades.

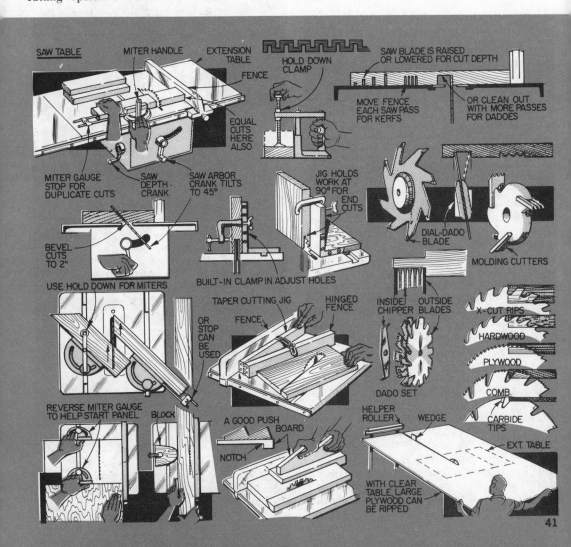

STATIONARY POWER TOOLS

The Radial Arm Saw

Originally conceived for industrial use, the many advantages of this popular tool can now be enjoyed by the average home craftsman.

A modern radial arm saw mounted on a steel stand with operating controls in front. Like the table saw, the diameter of its blade determines the maximum depth of cut.

A cousin to the bench saw is the **radial arm saw** preferred by some because of its unique design. With this saw, the work is held stationary; it is not pushed into the blade as with the bench saw. A long board can be clamped to the radial arm saw table while the motor and blade are pulled forward to do crosscutting. It can do the same work as the bench saw: ripping, crosscutting, mitering, dadoing, grooving and angle cutting. With attachments the radial arm saw can even be used for grinding and shaping. A special caution to observe when ripping on the radial saw is to always push the work into the blade from the direction marked on the guard; this is one of the few operations on the radial arm saw where the work is moved rather than the saw-motor. The ripping operation calls for the motor to be locked in place, parallel to the guide fence.

For mitering work, the saw-motor can be swung and locked at any desired angle. The cutting depth for all work is adjusted by raising the motor, up or down, by means of a crank. An advantage of the radial saw is that it gives the user a clear view of dadoing and cutting operations that are concealed by the bench saw. When cutting completely through the work, the blade cuts slightly into the table surface. This is of no consequence as the table is made of wood or hardboard and can be replaced when necessary.

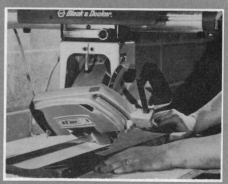

This isn't a bevel, or miter cut, as you might think. The operator is actually cutting a decorative concave in a board by tilting the blade, then pushing the board from right to left.

Crosscutting is easier to do with the radial arm saw than with the table saw because the work is held stationary. In fact, the work can be clamped in place; makes cutting safer.

A crosscutting operation. But, what's wrong here? The operator should have rolled up his sleeves before starting work. Well — he is not wearing a tie!

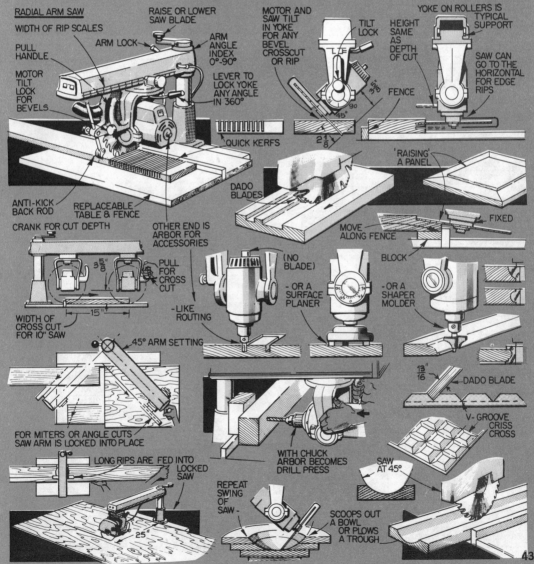

STATIONARY POWER TOOLS
The Band Saw

The **band saw** is one of the safest power tools in the workshop—and one of the most useful. It consists of a flexible steel blade over two large pulleys, and is an ideal tool for cutting curves in wood or plastic, for ripping long boards and for crosscutting, though crosscutting is limited by the distance between the post and the blade. Most band saws can cut wood up to six inches thick and can also cut wood at an angle by tilting the table to the desired degree. Band saw blades come in various widths. Use a narrow, ¼-inch blade for cutting sharp curves, and wide blades for straight-line cutting. When the blade becomes dull, discard it. It does not pay to resharpen as blades are relatively inexpensive.

The band saw is the ideal tool for cutting curves in wood; it can make cuts in wood that are as much as six inches thick. Mark the cut in pencil before starting; cut on waste side.

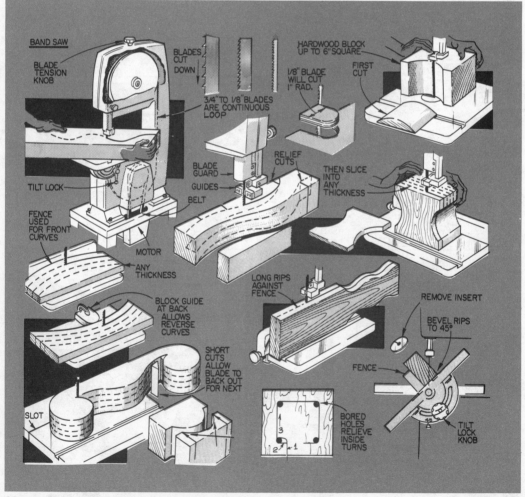

The Drill Press

The **drill press** is the second most important stationary power tool in the home workshop (after the bench saw or radial arm saw). Of course, it is used for drilling round holes, but it can also drill square holes, with a special mortising attachment consisting of a hollow chisel and drill bit. The nominal size of a drill press is the distance from the post to the middle of the chuck. Like many power tools, the drill press can also drill holes at an angle by tilting and locking the drill press table. When using the drill press, always place a block of wood under the work to be drilled. The block of wood will prevent the drill bit from splintering the work as it emerges at the bottom; it will also protect the drill press table. Always raise the drill press table as close as you can to the work piece. This will utilize the full length of the drill bit.

There are many attachments for the drill press: a countersink; a hole saw for large holes;

Speed of drill bit is determined by material and by diameter of hole to be cut. Large holes and tough steel require slow speeds; adjustment is made by shifting the belt.

a fly cutter for cutting out discs; a drum sander; files and rasps. The drill press should be adjusted to drill at low speed when drilling large holes or holes in tough metal and at fairly high speed when drilling small holes in wood and soft metals. Typical applications are shown.

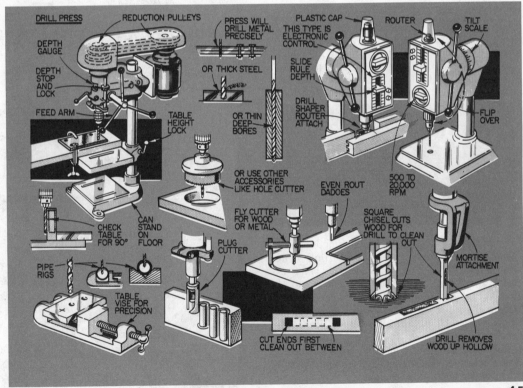

STATIONARY POWER TOOLS

The Sander

The combination **belt and disc sander** sands regular and irregular shapes to exact size or prior to finishing. The disc part of the sander is usually used for fast stock removal while the belt is employed for fine sanding. Fine to extra coarse grades are available for the belt as well as for the disc but, because the disc cuts across the grain, final finishing should always be done on the belt, with the grain. The table adjacent to the disk can be tilted for sanding miters to an exact 45-degree angle. It also has a miter-gauge slot for trimming work to fit. For sanding awkward or large pieces, the belt assembly can be raised to a vertical position, or to any position in-between. The rounded ends of the belt assembly are especially suitable for sanding interior curves. When sanding with the disc, hold the work so that the action of the disc forces the work against the table—*downwards*. When sanding flat areas on the belt, hold the work at a slight angle.

The Grinder

Grinders are designated by the diameter of their wheels. A good all-around grinder for the home workshop should have six-inch wheels. One stone should be coarse for fast stock removal.

Along with its nominal function, the **grinder** can also be used for buffing, wire brushing and polishing. However, its chief use is for sharpening tools to keep them at top efficiency. (A dull tool is more apt to cause an accident than a sharp tool, as a dull tool requires more muscle to use and thus causes slippage—and accidents.) Use the fine stone on the grinder for sharpening and the coarse stone for fast stock removal. Wear goggles and make sure the grinder guards are in place.

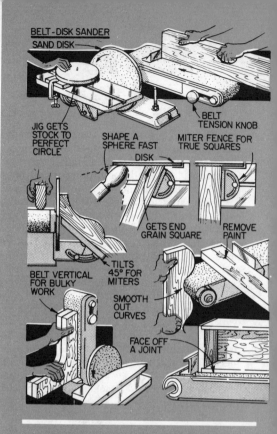

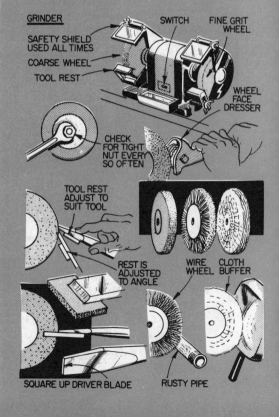

The Jig Saw

The **jig saw**, sometimes called the scroll saw, is a tool whose blade moves up and down. It can be used to cut intricate curves in wood and light metal. Its capacity is determined by the distance from the post to the blade. An important feature of the jig saw is its ability to make internal cuts. This is done by first drilling a pilot hole in the work and then inserting the blade through the hole. The jig saw table can be tilted to make angle cuts in work up to about two inches thick. There are many blades available for the jig saw; the narrower the blade, the smaller the radius it can cut. The blades are held at top and bottom, secured in a chuck by means of a setscrew. Heavy blades (called saber blades) are secured only at the bottom, as their stiffness does not require a support at the top.

The jig saw is used to cut external and internal curves—and straight lines—in wood or in metal with special metal-cutting blades. A pilot hole must be made to make internal cuts.

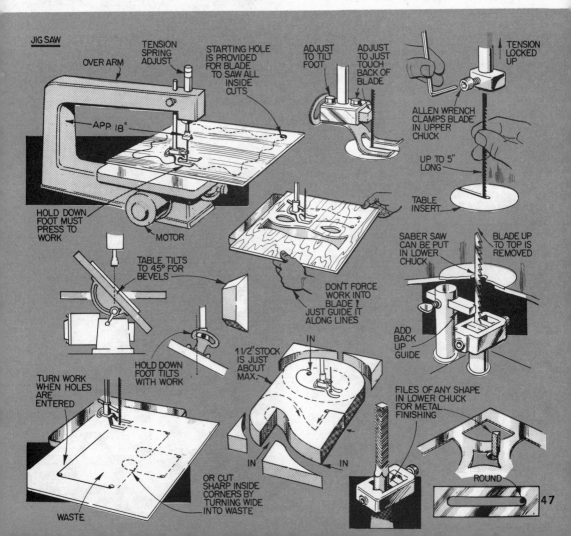

STATIONARY POWER TOOLS

The Shaper

The shaper is used to make fancy edges, can also produce molding.

The **shaper**, as its name implies, is a tool for cutting various shapes in wood. It can be used to make a molding from a length of ¾ x ¾-inch wood. It will make decorative beads along the edges of a panel. In effect it is a large-size version of the router, with greater capacity and much more power. Treat this tool with respect—its cutters revolve at extremely high speed. You can make all sorts of fancy decorative cuts by mounting combinations of various cutters and washers on the spindle.

Handle the shaper with care and respect as its cutters revolve at extremely high speed, some higher than 20,000 rpm. With a shaper you can design and cut molding to any contour you like.

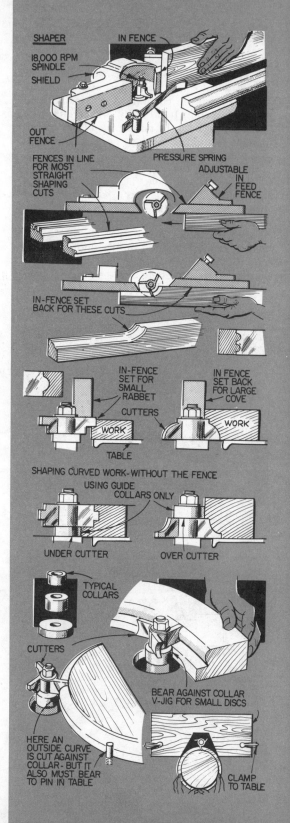

48

The Jointer

The **jointer**—its name refers to its ability to edge-plane two boards so that they will make a perfect joint—planes and edges lumber by means of a three-blade cutter that revolves at high speed between the front and rear tables. These table heights are adjustable. On some jointers, only the front table is adjustable; on better jointers, front and rear tables can be adjusted. The required depth of cut is made by lowering the front table to the desired depth. Like the shaper, this tool should be treated with respect and caution. The drawing shows how the table is adjusted.

Jointers come in two sizes, four-inch and six-inch. For average home use, the smaller job is fine, unless you believe you will have occasion to plane boards that are six inches wide.

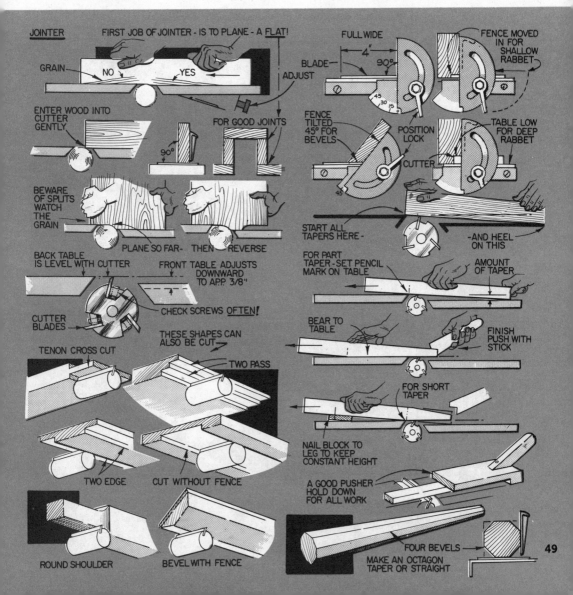

STATIONARY POWER TOOLS

The Lathe

The **lathe** has the ability to turn an oblong piece of wood into an absolutely symmetrical, tapered or non-tapered cylinder, complete with decorative ridges and indentations. It is the most fascinating tool in the workshop. However, its use is limited to turned work—and very often you can buy turned legs at little more than the cost of the raw wood.

If you do opt for a lathe, it will give you hours of pleasure as you watch the wood literally turning under your hands into a candlestick, a lamp base, a bowl, a tray, or even a baseball bat. The capacity of a lathe is limited by the distance between centers and the distance between the spindle center and the lathe bed. When you mount the work between the centers, it is called spindle turning; when the work is mounted on one center only—the one that is powered—it is called face-plate turning. Extra long pieces of work can be turned in the lathe by doing them in sections and then gluing them together, reinforced with dowels.

There are many more power tools, generally found only in industrial plants or in the shops of professional cabinetmakers. But the tools we have described so far will do 99 per cent of your work—no need to buy something you will use only once in a lifetime!

The wood-turning lathe is probably the most fascinating power tool in the home workshop. With it you can make a baseball bat, legs for tables and beds, vases, etc. In fact, any cylindrical object.

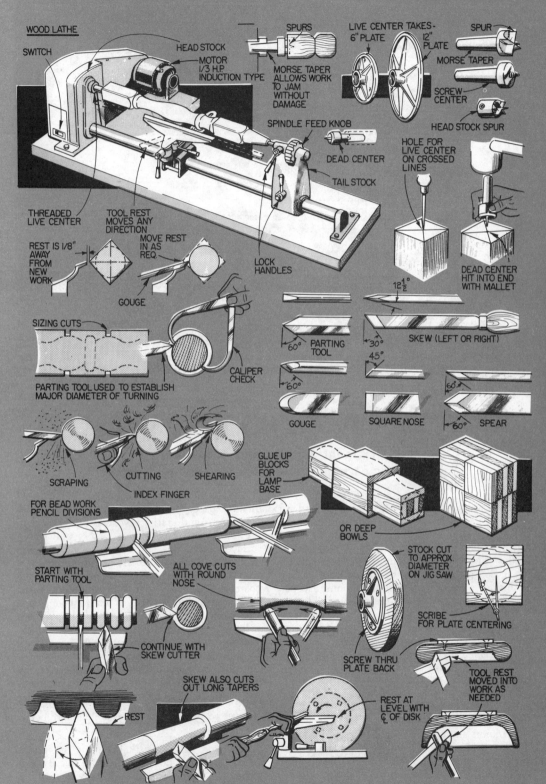

Lumber is trimmed to the best lengths that log length and imperfections will allow. Note the logs stacked in rear. The insert shows logs being cut to size by a band saw on a moving platform.

Lumber and Plywood

Some of pertinent information you should know about basic carpentry and woodworking materials.

The wood you buy from a lumber yard is available in two broad classifications: *select* and *common*. Select is used when appearance and finish are important and is divided into four grades. The best is B grade, sometimes called No. 1 clear; next is Better grade, or No. 2 clear. Both grades are devoid of imperfections, except for minute ones which are scarcely visible in No. 1 clear. C select grade has minor imperfections such as small knots; D select grade may have larger knots and imperfections which can be concealed by painting. Idaho white pine bears similar classifications: Supreme, Choice and Quality.

Common lumber has five classifications: No. 1 contains tight knots, few blemishes, can be stained or painted; No. 2 has somewhat larger knots and blemishes and is often used for flooring and paneling; No. 3 has some loose knots and may have some pitch blemishes, though still suitable for shelving; No. 4 is of low quality, used for subflooring, sheathing, concrete forms and crating; No. 5 is pretty close to firewood, with large holes, cracks and rough surfaces and sides; it is suitable only where strength and appearance are not important.

When you buy wood, you pay for it by the board foot. A board foot is the amount of wood in a piece of lumber that is one foot long, one foot wide, and an inch thick. If you buy a piece of lumber that is ten feet long, two inches thick, and one foot wide, you pay for 20 board feet. To calculate the number of board feet in a piece of lumber, multiply the length in feet by the nominal thickness and the width in inches and divide by 12 as follows:

$$\frac{10 \text{ feet} \times 2 \text{ inches} \times 12 \text{ inches}}{12} = 20 \text{ board feet}$$

A word about sizes. You are no doubt aware that a 2x4 does not measure two by four inches. A ruler will tell you that it really is

A saw mill planer cutting a profile on lumber to be used for siding. The same machine has already cut the tongue and groove on the opposite edge. This is the last stage in the planing operation.

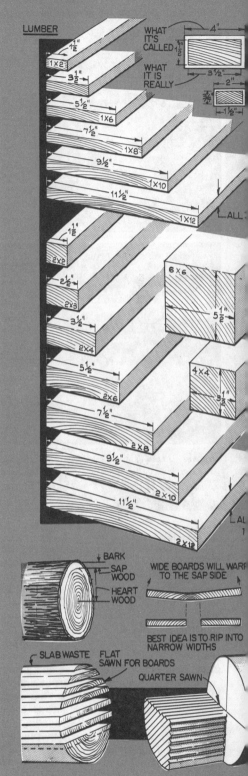

1½x3½ inches, sometimes a bit more and sometimes a bit less, depending upon the mill that supplied the lumber yard. The illustration at the right shows how the actual size of the lumber you buy differs from the nominal size. The length of a board, or 2x4 you purchase, is always as stated; it is not reduced by the cutting and sanding operation. Most plywood is sold in its actual size except for some ¼-inch paneling which is now being sold slightly less in thickness than a quarter-inch. This doesn't make too much difference once the paneling is up on the wall—but it does give the manufacturer more mileage out of a log.

Lumber, of course, is further classified into hardwoods and softwoods. The hardwoods, chiefly used for making fine furniture, are walnut, mahogany, oak, maple, cherry, rosewood and teak. The softwoods are pine, spruce, fir, hemlock and redwood.

Don't be afraid of working with hardwoods, treat them the same as softwoods. As long as your tools are sharp, the technique is the same.

Below: The first step in the operation of making plywood consists of removing the bark from the log. Next the log in mounted on a giant lathe and wood is peeled off like a ribbon from a spool.

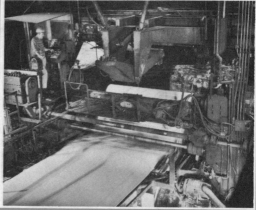

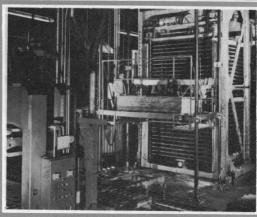

Above: Sheets of wood, after being cut to the right size are now ready for the gluing operation. Next a press binds the sheets of wood and glue into what now can be called plywood; sanding is next.

PLYWOOD

Plywood is another matter. The nominal thickness of a sheet of plywood is also its actual thickness. Plywood is made from an odd number of sheets of veneer, glued together with the grain of adjacent layers running at right angles. *Lumber core plywood*, used in furniture construction, has a single thick core of solid wood instead of several laminations. The most commonly available plywoods are the softwoods made from pine, fir and spruce and graded according to the quality of the outside veneers.

Most plywood is graded by the American Plywood Association with a stamp on the edges or on the back marked DFPA, standing for Department For Product Approval. Formerly these initials stood for Douglas Fir Plywood Association but inasmuch as not all plywood is made of Douglas fir, and the initials had a certain acceptance, a new name was found to fit the initials. Reproduced below is a typical inspection stamp. The two large capital letters refer to the grade of the veneer on front and back. A-A means that both sides are smooth, suitable for painting or for a less exacting natural finish. N would indicate suitability for a fine, natural finish with lacquer, shellac or varnish. B would mean that small repair plugs and tight knots are present, while C is for a face that has limited splits and somewhat larger knots; D indicates similar flaws, but larger than those permitted in the preceding grade.

Right below the veneer designation is a Group Number. This indicates the species group and relative strength. Group 1, the strongest, includes plywood made of sugar maple, birch, Western larch, loblolly pine, long and short leaf pine, and Douglas fir from Washington, California, Oregon, British Columbia and Alberta. Group 2 plywood is made of Douglas fir from Nevada, Utah and New Mexico, as well as cedar, Western hemlock, black maple, red pine and Sitka spruce. Group 3 plywood includes Alaska cedar, red alder, jack pine, lodgepole pine, Ponderosa pine, and red, white and black spruce. Aspen, paper birch, Western red cedar, Eastern hemlock, sugar pine and Engelmann spruce comprise Group 4. The last group, No. 5, consists of plywood made from balsam and poplar—the weakest of all the woods listed.

Note the word Exterior; it designates a plywood made with waterproof glue. If the word Interior is stamped on the plywood, a water-resistant glue was used in its manufacture; the wood is suitable for interior use, though not under extremely damp conditions such as in a bathroom. Sometimes you may find two additional numbers on the plywood such as 48/24. The first number indicates the maximum spacing between rafters for roof decking and the second number the maximum spacing between joints for subflooring. If the second number is a zero, such as 24/0, the plywood is not suitable for subflooring.

Lumber yards generally carry plywood in 4x8-foot sheets, in thicknesses from ¼ inch to ¾ inches. Large yards may carry, or can order for you, non-standard sheets. Because the corners of plywood sheets are often splintered and chipped, make allowance for this waste when ordering. Plywood can be cut, mortised, mitered and finished the same as solid stock wood. Always use a fine-tooth saw to cut plywood and, when using a hand saw, cut with the good side up to avoid splintering. When using a portable power saw, have the good side facing *down*. Pilot holes should always be made in plywood before using screws. Screws do not hold as well in the ends or edges of plywood as they do in the face. Before applying glue to plywood, roughen the edges to give the glue an extra "purchase." Exposed edges of plywood should be covered with molding or a filler strip, or packed with filler to hide the end grain. This does not apply to some imported plywood, made with 11 and 13 layers, that is actually so decorative it is left exposed.

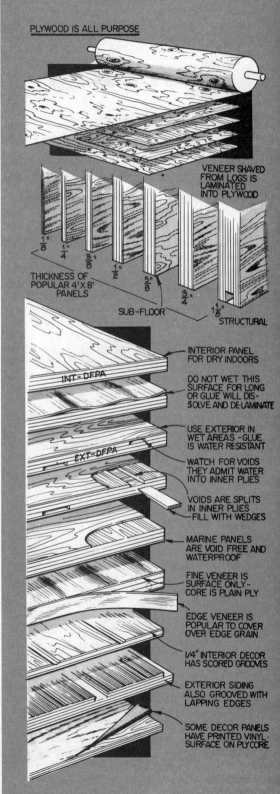

Wood Paneling

There is a wide variety of types available. Learn how to measure areas to be covered plus installation techniques for wood and plywood.

Installing paneling over existing plaster or directly on new studs is one of the most popular ways to finish a wall. While wood panels come in an almost infinite variety of wood species, when choosing your particular panel note the thickness. When the prefinished panel was first developed some 20 years ago the standard thickness was 1/4-inch, and while the majority of the higher priced panels are of this thickness, more and more lower priced panels are 5/32" or 3/16" thick. These thinner panels have the same beautiful surface as the 1/4" thick grades and are installed in much the same manner. However, some special procedures should be followed.

First, never install the thinner panels on bare studs or furring strips located more than 16" on centers.

(2) Always use an adhesive with the thinner panels because the grooves that are cut to simulate planks make the panel so thin at the nailing point that nails can easily pull through.

(3) Use the colored panel board nails with a ring shank. Where the top and bottom is covered by molding, use the large-faced 1 3/8" sheet rock nails.

The 4' x 8' panel is the most popular but 4' x 7' and 4' x 10' panels are also available. Of course, because of the greater length, 10 foot panels are substantially more expensive.

For easiest handling and to permit you to shift the panels as necessary to obtain a tight joint, cut each panel down so it is about 3/4" shorter than the ceiling height of the room. The extra space at the top or bottom will be neatly covered up by a base or ceiling molding.

Start working from one corner, and make certain the panel is absolutely vertical. If one edge fits into the corner and your level indicates the other edge is not perpendicular, plane down the corner edge accordingly. Work from the corner panel out and fit each panel before mounting it to the wall surface or studs. When you reach an electrical outlet, tape a piece of carbon paper over the outlet, (some mark the edge of the outlet box with chalk or even a dab of paint) and position the panel so the edges butt. Then place a block of wood over the general area where the outlet is located and rap it a few times with a hammer. This will transfer the outlet mark. Remove the panel, turn it face down and cut out for the outlet. If you allow yourself 1/8" around the mark, the space will be easily covered by the wall plate.

When securing a panel to a studded bare wall you have a few choices. (1) You can use an adhesive, (2) You can use pre-colored nails or (3) You can use a combination of both. We have found using both nails and adhesive produces a good job although for a PERFECT job with no visible nail heads only adhesive is used.

If you elect to use the colored panel board nails (they are far superior to the conventional wire brads), after the vertical joint is absolutely tight, *tack* (but do no drive home) the top to hold the panel in place. Then nail at 12" intervals in the groove above each stud working from top to bottom. Once the panel is flat against the studs, drive the top nails home and nail along the base. Where panels butt each other, drive 1" nails at a 45° angle so the heads are hidden in the groove. If necessary, you can also drive right through the face finish, sink the head slightly below the surface and then fill the hole with one of the many colored wood putties made just for the purpose.

If you use the popular adhesive method on bare studs, you still prefit and check the panel position by tacking in place. Then remove the panel and apply the adhesive to the studs. Run a single bead from top to bottom at the vertical joint where the panels will butt and an intermittent bead about 3" long every foot on the inside studs. Of course, you can run a continuous bead from top to bottom on all studs but the intermittent technique will save you quite a few adhesive dollars. It is not necessary to use adhesive along the top or base horizontal members; the 1 3/8" sheetrock nails do a superb job and are later covered up with molding.

Is it necessary to stud out an existing plastered wall before paneling? This is a good question, especially if you have a straight, structurally sound wallboard or plaster surface that you decide has had its share of paint or wallpaper. If this reflects your condition, fit each panel as previously described and use the adhesive method if the base material won't peel off. If necessary, you can drive nails in the V-groove (actually

A keyhole saw is used to cut an opening for the switch box shown near the man's head. A piece of carbon paper taped over the switch box and a blow with your fist will transfer exact area to be cut.

After applying the recommended panel adhesive, drive a half-dozen nails an inch or so away from the ceiling. The nails will form a sort of rough hinge for the next step illustrated in photo below.

Pull the panel away from the wall about four in. and keep it in place with a block of wood. Allow the adhesive to set for a few minute, until it becomes tacky, remove the block and push panel in place.

You can also nail panels in place with brads and a nail set, or by means of this device called a Whammer, which shoots small nails into the panels. Drive brads along the vertical lines of the panels.

they are not V-shaped any more) right into the plaster.

With all your panels in place finish up with one of the prefinished baseboards, cove or corner moldings. These are nailed in place, using colored 1 3/8" panel board nails with a ringed shank. They do the best job and are nearly invisible.

CALCULATING PANEL NEEDS

A simple way to calculate how much paneling you need for a room is to first draw the outline of the room on graph paper. Let's assume your room measures 16x20 feet. Each square on the paper will represent two square feet. The room will then measure eight squares by ten squares as shown in your sketch. Now measure the height of the ceiling and draw the sides of the room, treating them like the dropped down sides of a box as shown. Indicate the location of the doors, windows, and fireplace, if any. If there are no doors or windows, you merely add up the perimeter to get a total of 72 feet. Because each plywood panel is four feet wide, divide four into 72 to get the number of panels required, 18. In the sketch, there are two doors and three windows. Make an allowance of ½ panel for each window (1½ panels), and ¾ panel for each door (1½ panels), for a total of 3 panels to be subtracted from 18, or 15 panels to buy. If in doubt, buy an extra panel.

SOLID WOOD PANELING

You can also panel with solid wood, available in ¾-inch thickness and random widths. There are both softwoods and hardwoods, and can even be bought tinted if you prefer color. Three popular styles are the butt joint, shiplap joint and contemporary joint.

Solid wood paneling is generally applied over furring strips, either horizontally or vertically. It can also be applied horizontally over conventional 2x4 studding. The furring strips—1x2 or 1x3—are nailed to the walls at 16-inch intervals, center to center. If the walls are uneven, as may be the case in a basement, the strips should be shimmed out with thin strips of wood. *Caution:* If the room to be paneled is inclined to be damp, or if you are planning to panel over masonry, coat the walls before you begin with a waterproofing material. It is also a good idea to back-prime the panels with shellac to stop dampness penetration from the rear. Good ventilation is an important requisite when paneling a basement room. The use of furring strips allows air circulation behind the panels. For this reason, panels in basement rooms should never extend completely to the ceiling or floor. The space at top and bottom is covered with molding. You can buy crown and baseboard molding with inconspicuous slits for ventilation purposes.

Of the solid wood used for paneling, pine is the most popular. On a square foot basis it is usually more expensive than the popular 4 x 8 ft. prefinished panels. 1. Preparing a room for solid panels is no different than if you used prefinished panels with the one exception that you need not use any vertical furring strips. Here 1 x 4 in. strips are being nailed in place. 2. Fitting around corners is easier because you are working with narrow boards rather than 4 x 8 ft. panels. 3. This knotty pine room takes advantage of the many different types of molding that are available for ceiling trim and to create a special framing effect. 4. Rustic den installation with cabinet doors made from the same paneling used on the walls.

INSTALLING THOSE LOVELY PLYWOOD PANELS YOU CAN BUY TODAY

- STAND PANELS ALONG WALL FIRST TO CHECK GRAIN AND GROOVE MATCH
- POSITION FIRST FULL PANELS IN PROMINENT AREA – GET THEM PLUMB THEN WORK TOWARD CORNERS
- AVOID CORNER STARTS IT COULD BE OUT OF PLUMB
- PLAN TO WORK FROM WINDOWS OR DOOR FRAMES ALSO – LETTING ODD REMNANTS RUN TO WALL
- NAIL LIGHTLY AT TOP FIRST TO LET PANEL HINGE OUT AT BOTTOM
- REMOVE LOOSE PAPER
- IF WALL IS CLEAN, FLAT, DURABLE YOU CAN STICK THE PANELS RIGHT TO IT IN ADHESIVE
- IF WALL IS OLD OR UNEVEN – YOU MUST NAIL UP 1" X 2" LATERAL FURRING TO SUPPORT PANELS
- FIND WALL STUDS FOR NAILING
- SPACE 16"
- ADD VERTICAL 1" X 3"S AT PLACES WHERE PANEL BUTTS
- SHIM OUT HOLLOWS
- NAIL INTO WALL OR CEILING OR JAM WITH WEDGES
- ADHESIVE CAN BE USED IN ANY SYSTEM SHOWN
- POLYETHYLENE BEHIND AS VAPOR BLOC
- ON BASEMENT OR OTHER MASONRY WALLS – ERECT NEW SURFACE OF 2" OR 3" STUDS
- 1" X 2" CAN BE NAILED OVER STUDS – ESPECIALLY IF PANELS ARE 3/16" OR LESS
- 2" X 4" ARE GOOD STIFFENERS
- INSULATE HERE IF NEEDED
- NEW WALL WORK IS STANDARD 16" STUD SPACING – TAKING CARE TO GET PROPER 48" PANEL EDGE SUPPORT
- ODD
- OUTSIDE AND INSIDE CORNER
- FURRING
- MITER
- OR LAP
- PLAIN BUTT
- OR COVE
- ALLOW 1/4"
- CEILING COVE
- NEW BASE MOLDING
- NAIL INTO GROOVES
- PANELS
- FURRING
- CASING
- NAILING AT BUTTS
- PANEL IN ADHESIVE
- CAP MOLD
- DOOR AND WINDOW TRIM

Wallboard, Hardboard and Plastic Laminates

Special techniques needed for the installation of these various wall coverings are shown and explained.

Prefinished pegboard with either 1/8 or 1/4 in. holes is ideal for the workshop and kitchen wherever storage wall space is needed. It comes in two types; tempered and regular.

There are other building materials that you should be familiar with. One of the most common is *wallboard*, a product used in so-called dry wall construction. This material consists of a fire-resistant mineral (gypsum) covered on both sides with heavy brown or gray paper. A few firms produce decorated wallboard to simulate wood grain and other textures. It comes in sheets four feet wide and up to 16 feet in length; thicknesses are 3/8-inch, ½-inch and 5/8-inch. Most wallboard panels are slightly tapered at the edges to allow for the application of a special joint compound and tape to hide the seams between panels. Some other wallboard edges are shown.

The tools you need to work with wallboard are simple and few: a hammer, a ruler, a straightedge, a trimming knife and some sandpaper. To apply the joint compound you will need a trowel and special perforated tape. To cut wallboard to size, first score through the face paper and into the core slightly, using the knife and the straightedge. Support the wallboard along the line of cut and break it by bending, or snapping it *away* from the scored line. Trim away the partially broken paper on the hinged side of the cut and use the sandpaper to smooth the edge.

When applying wallboard to exposed 2x4 studding, it is advisable to use ring-type nails to avoid possible future nail popping. Nails should be 1¼, 1½, or 1-5/8 inches long, depending upon the wallboard thickness. The nails should be staggered on each side of the joint. A trick used by many professional installers is to stagger *two* nails on each side of the joint, as a precaution against later loosening and popping out. Nails should be slightly set below the surface of the wallboard with the last hammer blow. This slight "dimple" allows the wallboard compound to hide the nail head. Use a putty knife to apply the compound over the nail heads and over the seam and apply the tape before the compound has a chance to set. Feather out the edges beyond the tape area. Allow this first coat to dry, usually overnight, and then apply a second coat, this time feathering it out several inches wider than the first coat. When the second coat has dried thoroughly, sand the surface carefully with a medium-grade sandpaper. Apply a primer coat before painting or shellac if the wall is to be papered.

Hardboard comes in standard 4x8-foot sheets, ¼ and 1/8-inch thick, though you can get them in six- and ten-foot lengths on special order. They are also available in 16-inch planks

This wall is covered with Presidio, a hardboard reproduction by Masonite of an 18th century Spanish wall surface. Each 4x8-ft. panel appears to be made of 18 individual hand-carved wooden blocks.

for use as wall paneling. Perforated hardboard is popular for display backgrounds, and tool and utensil racks. Many hardboards come in a vast array of colors, designs and patterns. Finishes are baked on in the better grades, and rolled or sprayed in less expensive grades. Some of these hardboards closely resemble finished hardwoods and marble. Hardboards are applied with cement, cement and nails, nails only or clips, to solid walls, furring strips or wall studs.

Plastic laminates, sold under such trade names as *Formica* and *Micarta*, are also used for walls, especially in kitchens and bathrooms. They too are available in many patterns, simulated wood grains and solid colors. Laminates can only be applied to a perfectly flat, level surface. A smooth backing of plywood or hardboard is satisfactory. Since clamping is impossible in most installations, contact cement is nearly always used in applying laminates.

Sheets of plastic laminate measure up to ten feet long and four feet wide. They can be cut to size with a fine-tooth saw. Cut at a slight angle with the good side up, except when using a portable power saw—then the good side is down. Roughen the surface to be covered with coarse sandpaper and apply a coat of contact cement. (Painted and varnished surfaces should

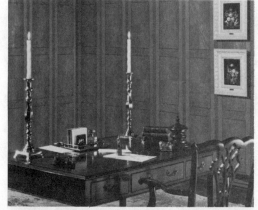

Two more examples of elegant wall coverings. Such wall coverings characterize the trend toward maintainance-free housekeeping. A swipe with a damp cloth is all the care many of these modern wall coverings require. Matching moldings are available.

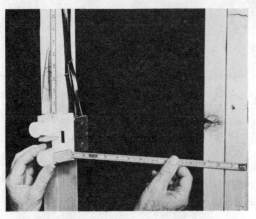

The most difficul part of paneling is cutting out for the electrical outlets. Some craftsmen make a paper template but this new aid called Panel Pal by Toolco, makes the job easier.

The flanged base of Panel Pal fits into the outlet box (left) and then the horizontal and vertical rules are pulled out and locked. The tapes and base are then transferred to the panel.

With a regular calking gun, apply a continuous bead of adhesive on the vertical furring strips where panels join. Apply intermittent (3 in. bead with 6 in. open space) on horizontal strips.

With the panel adhesive on all horizontal and vertical furring strips, rest the edge of the panel on the floor and carefully nudge into position. Move it as necessary so it is absolutely vertical.

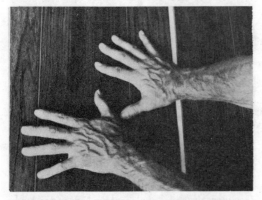

Unlike contact cement, panel adhesive does not "grab" immediately. You can easily move the panel in any direction to get the vertical joint absolutely flush against the adjacent panel.

Before the adhesive sets, the panels may tend to "bow out" from the wall. A square block of wood covered by a scrap of carpeting and a few firm raps with a hammer will push the area down.

After the vertical joints are tight, tack the upper and lower edge so the panel will not shift. Do not drive the nails home; keep them in an area that will eventually be covered by the molding.

If you elect not to use molding, remove the nails that were used to position the panel and prevent it from shifting. A smooth piece of plywood, under head will prevent damaging the surface.

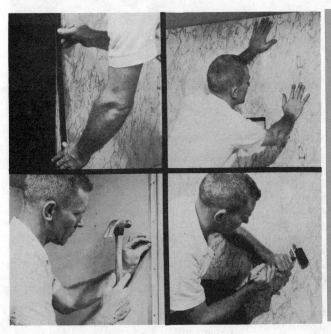
Waterproof marble-faced panels are ideal for bath and shower walls. The important point to remember is that all panels should be prefitted first, and then cemented in place. Make them about 3/16 in. smaller than the wall area. 1. Always work in one direction—left to right or right to left. Nail the aluminum or plastic molding to a corner. Check to see that the panel fits. Then apply the mastic as recommended by the manufacturer. 2. Smooth out the panel on the wall, pressing down any areas that tend to pop up. 3. Use a roller to smooth out the mastic. 4. The corners are handled with a special piece of molding.

be sanded to the bare wood before application of the cement.) Apply another coat of contact cement to the laminate. Let both coats dry for about 15 minutes. Because contact cement "grabs" immediately, a "slip sheet" of paper is required when you position the laminate. Lay the paper over the surface to be laminated and then place the laminate over the paper. When you are certain that the laminate is in the correct position, lift one edge of the laminate, tear off a few inches of the paper and carefully lower the laminate so it grabs. Now lift the opposite end, remove the remaining paper and lower the laminate into position. Press down over the entire surface and rub with a towel to make sure good contact is achieved, especially at corners, edges and seams. It is best to back all laminate-covered surfaces whose rear is exposed to the air with a low-priced backing laminate to prevent warping. However, a countertop or a table top which is securely attached to a framework need not be covered.

A fairly well-equipped workshop with all of the tools, and even the work benches, mounted on casters. It occupies one-quarter of a 30x50-ft. basement. Small hardware is stored in steel cabinets at the top.

Planning Your Workshop

Whether in basement or garage, a place to work and where your tools are readily available will make your projects easier to handle.

Most home workshops are located in the basement, though if you have a fairly large garage, there is no reason why you cannot devote part of the space to shop use. In this case you will be probably be limited only to a workbench, plus one or two stationary power tools such as a circular saw and a drill press. One great advantage of the garage workshop is that you will be working on grade, with the sun streaming in during fair weather.

But if you have a basement, that is the ideal place for a workshop. The family will be away from the noise, dust and chips as you are busy making the shelves, cabinets and space-saving built-ins that your wife has been urging you for years to finish.

How much room do you need? Very little. In fact, an area measuring only 5x15 feet will do nicely. Within this space you can find room for a 2x5-foot workbench, a circular saw (or a radial arm saw), a drill press and a band saw. If possible, select a corner of the basement so that you will have the alcove formed by the two walls. You should have at least three droplights for illumination, each with a 100-watt lamp—or a row of 40-watt fluorescent lamps if you prefer. A floor plan for this workshop is shown; the construction details for the workbench are described on page 68. The three stationary power tools will take care of most of your building requirements; these will be supplemented by portable power tools.

The wall area immediately above the workbench is the place where you will want to store your hand tools. The best way to do this is to mount a sheet of ¾-inch plywood, about as long as the workbench and two feet wide. On this board install a series of hooks and nails for hanging the tools. This is much better than having to rummage through a drawer in order to find a screwdriver or a hammer. With a little ingenuity and imagination, you can find a way to hang every tool you own. In addition to hooks and nails for supporting tools, you can sometimes drill a hole through the handle of a tool to hang it, or pass a string through the hole for hanging. About the only tool that does not lend itself to hanging is the plane. Best to make a small shelf for it on the tool board. Store the plane on its side, not upright. This way you will be less apt to dull the blade.

A workshop set up with the larger power tools in center, workbenches next to the walls and plenty of outlets in a power strip running along the wall. Power outlets are on a separate circuit from lights.

Another method of storing screwdrivers, awls, chisels and similar tools having a handle wider than the business end is to drill a series of various-sized holes in a four-inch wide board and screw the board to one side of the workbench. When you are finished with the tool, just drop it into any hole that will pass the tool —but not the handle.

The space alongside the workbench can be used for a series of shelves for storing paints, nails, brushes, miscellaneous hardware and a supply of wood. If the joists of the basement are exposed, do not overlook this area for storage. Spaces between the joists are especially handy for storing long dowels, pipes, molding, narrow boards and similar items. Nail a couple of slats across the joists to support such items. Or nail hardware cloth between the joists for storage purposes—the hardware cloth can hold small items and at the same time keep these cans, boxes of nails, etc. readily visible.

The workbench should be equipped with a vise. The matter of which end of the workbench to attach the vise is one of personal preference —the right or left side should have no bearing on your political leaning. Better yet, mount a

Install casters so you can move your power tools. Use casters that have rubber wheels that tend to take a "set" to keep the tool stand stationary, yet an extra hard pull or push will get them rolling.

Below: Best to save one part of a workbench for a drill press and a grinder. When building your own workbench, make it so that it is at a comfortable height, generally 36-in. high; allow for toes.

Top: A handy rack for tools can be made by drilling a series of holes in a piece of ¾-in. plywood and mounting it at the end of the workbench. It makes a good storage spot for screwdrivers, chisels, etc.

Below: These two mobile storage racks also serve as "room dividers" for a basement workshop, setting off the "play" area from the workshop. Note that shelves in both racks are of varying heights.

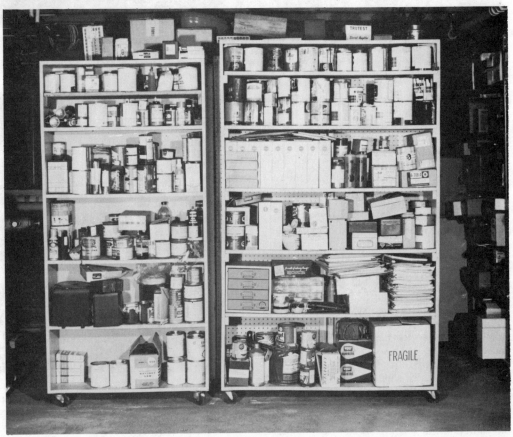

woodworking vise on one end and a metalworking vise on the other. As shown in the photo, the bench we constructed has a metal-working vise mounted on the right side of the bench. In the drawing is a floor plan of a workshop measuring only 5x15 feet. Note the placement of the power tools. Because the width of the workshop was limited, the circular saw was placed against the wall; however, it is mounted on casters so that it can be moved away from the wall for cutting large sheets of plywood.

Here is a floor plan for a somewhat larger workshop which includes additional power tools —a jointer, a jig saw and a sander. Since a jointer is usually used in combination with the saw, it should be placed in the immediate vicinity of the saw, whether a radial arm or a bench type.

It is a good idea to equip all stationary power tools with casters so they can be moved out of the way when necessary. You can get casters that have a locking brake on them to prevent rolling.

Note the placement of the saw in the center of the shop—an ideal spot if you have the room since this will give you plenty of clearance for ripping and crosscutting large panels. Even so, mount the saw on casters.

As mentioned before, the workshop should be adequately illuminated, and the power source for the various tools should be on a separate circuit. There is nothing so dangerous— and frightening—as to blow a fuse while sawing a particularly tough piece of wood and have all the lights in the shop go out as well.

Many home workshop fans have two workbenches—the first, with a vise, for doing the "work"; the second used for tools such as a grinder, a buffer and a drill press. While these tools are available as floor models, it is generally more convenient to buy the bench-mounted type so the floor can be utilized for those stationary power tools that cannot be bench-mounted.

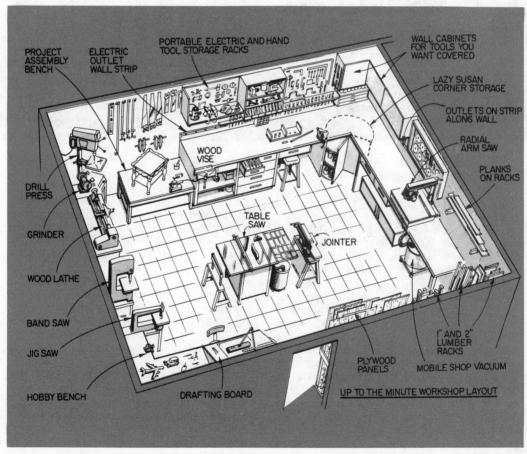

UP TO THE MINUTE WORKSHOP LAYOUT

Building a Workbench

Your first project, of course, should be a sturdy workbench.

Use C clamps around the edges of the top and buckets filled with sand to apply pressure to the plywood and hardboard top while the glue sets. Remove after 24 hours and trim all the edges flush.

The finished workbench, made out of 2x4s, plywood and hardboard, can be completed in 24 hours. Expedite the job by having the lumber yard cut a 4-ft.-wide panel down the middle to deliver two 24-in. wide lengths.

The bench has two identical frameworks, one for the top and a second for the bottom. The joints at each end, dadoed, glued, and bolted are very strong.

A good deal of the preparatory work for do-it-yourself projects around the house is done on the workbench. Here's a bench that is easy to build, using standard 2x4s, ¾-inch plywood and a ¼-inch hardboard top.

First cut the four 2x4 legs, each 34 inches long. Next cut 2x4s for the upper and lower framework, to the lengths indicated in the drawing. Notch each of the four long lengths to accept the cross pieces. The notches, or dadoes, should be about ½-inch deep and just wide enough to fit the 2x4 cross piece. Inasmuch as 2x4s vary slightly in dimension, carefully check the thickness of the 2x4s you are using. The notch should be about three inches from each end as shown in the drawing. Next apply some white glue to the inside of the notch. Drive the two cross pieces into the notches. They should be a snug fit. Drill pilot holes, for ¼-inch lag bolts. Inasmuch as you will be driving the bolts into the end grain of the cross pieces, use lag bolts at least four inches long to assure a good purchase. Assemble both frameworks. You will now have two oblong frames, 60 inches long and 20 inches wide.

With the upper framework resting on the floor, clamp the legs (using C clamps) in each corner, then drill clearance holes through legs and frame for ¼ x 4-inch carriage bolts. Apply white glue where the surfaces will meet, insert bolts in the holes, add a washer and nut to the end of each and tighten securely. Remove the clamps. Now gently ease the second frame over the legs so that it is about six inches from the ends of the legs. Clamp it in place, drill clearance holes as before and insert carriage bolts, washers and nuts to finish the lower part of the workbench.

The top is made of two layers of ¾-inch plywood, glued together and faced with a sheet of ¼-inch tempered hardboard. Buy a single 4x8-foot panel of utility grade ¾-inch plywood. If the lumber yard will sell you a shorter length, such as 4x6 feet, fine. Cut the sheet in half the

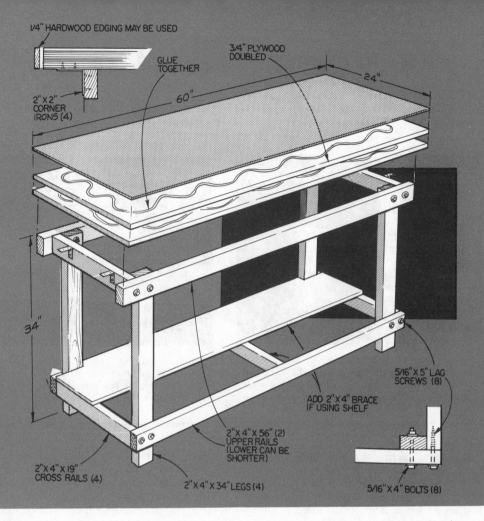

long way, then cut each piece so that it is 60 inches long. You will now have two 24x60-inch pieces of plywood. Coat the top of one piece with a layer of white glue, squeezing the glue out of the dispenser and using a brush to distribute it over the panel. Pay special attention to corners and edges. Carefully place the second panel over the glued panel, line up the edges, and nail the two panels together. Use annular ring nails—they hold better. The nail heads will be hidden by the hardboard top.

Cut a sheet of ¼-inch hardboard to 24x60 inches. Apply white glue to the top side of the glued-up plywood—the side with the nail heads showing. Use a brush as before. Set the hardboard in place. This time you will have to use clamps or weights to apply pressure while the glue dries. Do not clamp directly to the hardboard; use scrap wood to spread the pressure and to avoid marking the surface. You won't be able to clamp the center of the workbench top so use weights for pressure. A few buckets filled with sand or even water will do. Allow the entire assembly to dry overnight. Remove the clamps and weights and install the laminated top to the workbench framework. Its own weight will help to hold it in place. To prevent shifting, use an angle iron at each leg to secure the top.

PARTS LIST (dimensions in inches)

Two	¾x24x60	plywood (for top)
One	¼x24x60	tempered hardboard
Four	2x4x34	legs
Four	2x4x20	crosspieces
Four	2x4x56	frame members
Eight	5/16x4	lag bolts
Eight	5/16x4	roundhead bolts, nuts
Sixteen	3/8-inch	washers
Two	16-ounce	containers of white glue

Splices and Joints

Special joints do special jobs. The best joint is the most simple joint that will serve the purpose. The most useful are described.

Electrichisel, made by Stanley, is chucked into an electric drill and is used for dado work and mortising. The revolving cutting edges slice through wood the same way a hand chisel does.

The tenon, cut by hand by means of a back saw, is being fitted into the mortise cut by the Electrichisel. Break the leading edges of the tenon by light sanding so it will slip into mortise easily.

When you have familiarized yourself with the tools described earlier in this book, you will want to take a look at some of the many splices and joints every home carpenter should know, and know when and how to use them.

A **splice** is a means of connecting two lengths of wood, end to end, to run in the same line. A splice can be made with a "fishplate" nailed to one or both sides of the connecting members; with a half-lap joint; a splayed lap; a scarf joint; a bolted joint; or with a V-splice joint (see drawings).

Joints are used to connect wood at an angle. Hardware can be used to make a joint. For example, a corner brace, a T plate or a corrugated fastener may be used to make corner joints, such as in wooden window screens.

Full- and half-lap joints are used for fitting cross rails flush to their meeting members. As shown in the drawings, both are simple to make. Use a back saw and sharp chisel to make the required cuts.

The plain and stopped dado are more sophisticated methods of joint-making. In the plain dado, the cut-out area extends across the complete width of the work. In the stopped dado, the cut stops short of the end of the work. This results in a clean-looking joint and is used quite often when appearance is a factor. The work that is to fit into the stopped dado is cut away, as shown in the drawing.

The mortise and tenon is a still more sophisticated joint. It is the strongest of the T joints and is generally used in heavy framing work. The thickness of the tenon (the tongue) should never be less than one-third the thickness of the stock from which it is cut. A variation of the mortise and tenon joint is the stub tenon where the tenon goes only part way into the mortise. A double tenon is just what its name implies, as shown in the drawing.

The dovetail joint is the strongest of all corner joints. Careful marking and sharp tools are important requisites for its construction. Dovetail joints fall into several categories. The most common—and simplest—is the through dovetail. A bit more complicated is the lap dovetail, often used in bookcase and drawer construction. The double lap dovetail shows very little end grain but demands great care in its construction. The secret, or miter dovetail, is often used for high quality cabinet work. The knife-edge miter requires great care in its execution. See the drawings for details.

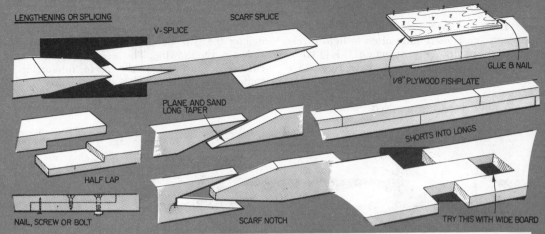
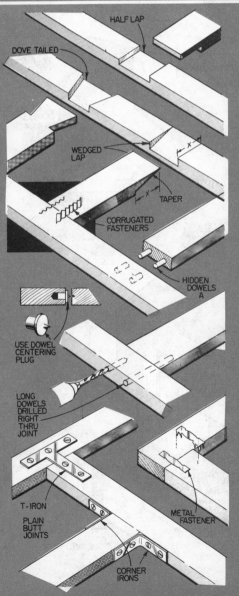
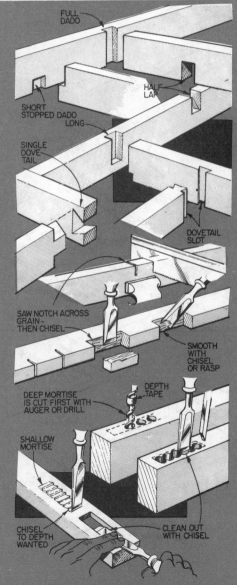

The box joint is a sort of simplified dovetail joint and is used quite often in light furniture work and drawer construction. It is of fairly simple construction and if you have access to a bench saw, the required cuts can be made in minutes.

A miter joint is assembled by one of four methods. The simplest—and weakest—is by nailing the two meeting members together as shown in the drawing. The second method is to cut two or more slots in the top for reinforcing splines. A third method is to cut along the width of each piece to accept a single long spline as shown. The last method is to reinforce

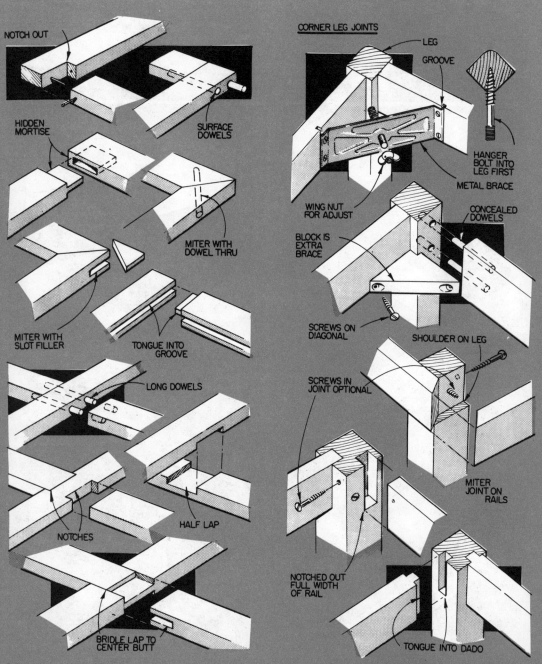

the miter joint by means of dowels. To assure accuracy in the placement of the dowels, temporarily drive a couple of brads into one face. Cut them off, not quite flush, and press the two pieces together so that the brads will leave their marks on the other face. Remove the brads and drill holes for the dowels. Result: a perfect fit!

Cross-over joints. The simplest cross-over or X joint is the plain overlap. Glue, nails or screws secure the joint. A neater version is the cross-lap joint made by cutting a half-lap in each piece of wood. A third way of making a cross joint is by means of two or three dowels as shown in the drawing. A mitered bridle joint takes a bit more care in its construction, but it is strong and somewhat decorative. A fifth way of making a cross joint is by means of a mortise and tenons. Use white glue to secure the joint.

Edge-to-edge joints can be made in several ways. The easiest and the least strong is by gluing the work together using bar or pipe clamps as shown. Dowels make a stronger joint, but require accuracy in preparation. A splined joint is the strongest joint of all for edge-to-edge work. However, it does require a power saw to cut the required grooves accurately. The grooves should be slightly deeper than the spline to allow for glue "expansion."

Lengthening joints. There are six common methods of making a long piece of wood out of two short pieces. They are the lapped joint, the splayed lap, the scarf joint, the V splice, and the bolted joint using fishplates or joining plates as they are sometimes called.

Three-way joints. Attaching legs (for example, in the construction of a workbench) calls for three-way joints. The drawings show five ways to attach legs. No. 1 is by means of a metal corner brace, a wing nut and a hanger bolt. These corner braces are made commercially; you can buy them in any hardware store. Note the slots cut in the side members to accept the curved edges of the metal brace. No. 2 shows construction and assembly by means of dowels. Dowel placement should be staggered to prevent their meeting in the middle of the leg. Use two or more dowels for each leg. Nos. 3 and 4 are very similar in construction, except for the placement of the leg. No. 3 has the leg on the inside while No. 4 has the leg on the outside. No. 5 is a haunched and mitered mortise and tenon joint; it is the strongest of the leg joints. When making this joint, leave a quarter-inch or so of waste at the top of the leg and trim it off with a plane and rasp after the glue has dried.

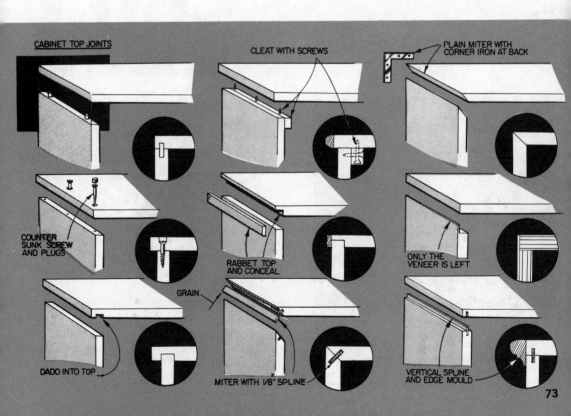

Coped joints. This joint is used in combination with a miter joint when installing ceiling and floor molding. It is not necessary to miter all four corners of this molding. The professional will first put up two lengths of molding, at *opposite* sides of the room, butting them snug against the wall (see drawing). The other two pieces of molding are then mitered and coped to fit. The advantage of a coped joint is that it will hide any irregularities due to walls not meeting at an exact right angle.

If the molding is flat on the back, you can use a short piece of scrap molding as a pattern. Trace the contour as shown in the drawing on the flat side. Tape the front (to avoid splintering) and cut through the molding with a coping saw. If the molding is irregular on the back, which is usually the case with crown molding used at the ceiling, you will have to use another technique. First cut the molding at a 45° miter. Make sure you have measured carefully. Then cut away excess wood following the outline of the miter. As you cut, tilt the coping saw slightly to the long side of the molding instead of cutting straight across at a 90° angle. This way you will gain a bit of extra clearance.

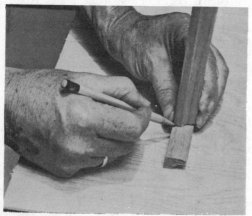

Molding can be fitted by coping instead of mitering. The first step is to trace the outline of molding to be cut, on the back, using a piece of scrap of same pattern as a guide. Use a sharp pencil.

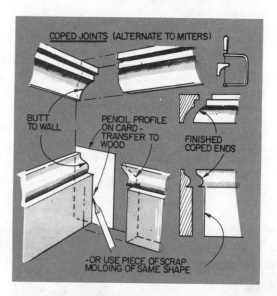

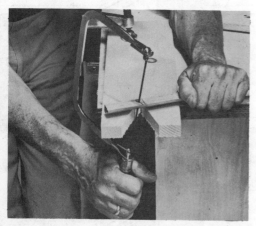

Next, cut the molding along the penciled outline with a coping saw. Tape the front of the molding to prevent possible splintering. Note the use of a wedge opening in wood block for a support.

Try the two pieces for fit. During the cutting operation, the saw should be tilted slightly instead of cutting straight across the molding in order to gain extra clearance. Practice on scraps first.

THE MITER BOX

In all coping and mitering operations, a miter box is essential. The miter box practically guarantees an exact 45° or 90° cut. You can buy an inexpensive miter box of wood, or a slightly more expensive miter box made out of steel and adjustable to in-between angles. Or you can make your own miter box—it is quite simple. Here's how: nail together three pieces of wood to form a long U-shaped channel. The inside of the U can be four or five inches wide—exact size is not important as long as the molding can fit into it. Use a try square to mark two lines perpendicular to the parallel uprights as shown. Now measure the *outside* width of the channel. Let's assume it is five inches wide. Lay off five inches from each line you drew, measuring from the outside as indicated. Draw a connecting line. In effect you have laid out a square and its diagonal—and the diagonal of a square is always exactly 45°. Draw another diagonal, crossing the first one. Now use the back saw and carefully cut across the box into each diagonal line. When you reach the bottom of the channel, stop. You can now use the box to cut molding at a left or right 45° angle. Complete the miter box by making a 90° cut (but not on the perpendicular lines you first made). The 90° cut is used to make true square cuts across molding.

This metal frame miter box made by Stanley is designed to fit any standard back saw. It is used for right and left miters.

When using the miter box, nail the molding to the bottom of the box on the waste side to prevent slipping as you cut.

Another good wrinkle is to protect the bottom of the miter box with a piece of scrap wood as this operator is doing.

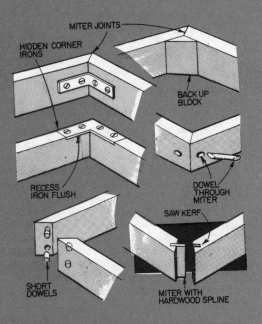

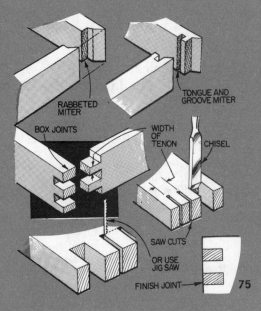

Nailing a plywood sub-floor to the joists; close nailing at this time will assure a squeak-free floor in the future.

House Framing

A knowledge of framing is absolutely necessary whether you intend to build a house, add an extension, or even an extra room.

The framing of a house depends upon its style—split level, one story, two stories, gable roof, hip roof, etc. The most common type of framing is *platform framing* (sometimes called western style). Another type—seldom used nowadays—is *balloon framing* in which the studs are continuous from foundation to the roof rafters.

In platform framing, the first floor is built as a platform on top of the foundation—hence its name. The first step is the installation of the sill, a 2x6-inch timber that is anchored in place by bolts previously set at intervals in the foundation. This sill extends all around the perimeter of the house. On top of the sill are placed the joists—supports for the floor. Joists always run the short way of the house. If your house is 30x60 feet, the joists will span the 30-foot dimension, every 16 inches along the 60-foot length. But, joists that long need additional support at or about the middle. Such support is provided by a girder, sometimes called a beam, that rests at each end on the foundation walls and is supported every ten feet or so along its length by posts or columns. This girder may consist of a solid timber, a built-up beam or a

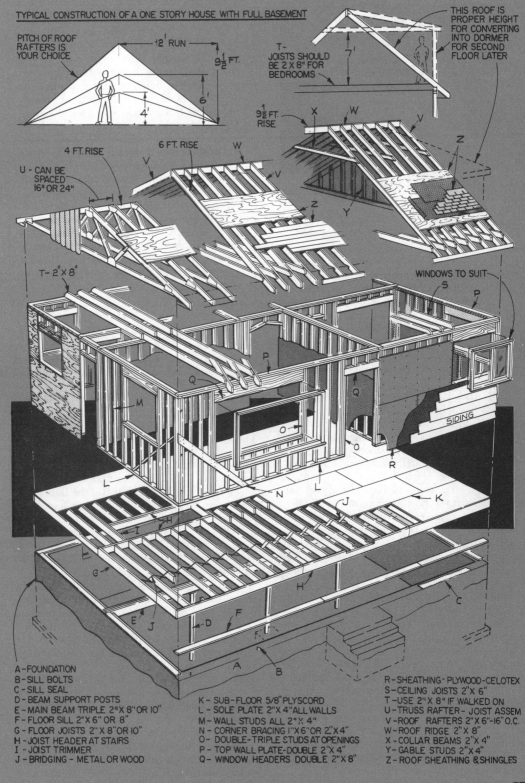

Special adhesives can also be used to secure the sub-floor to the joists. A heavy bead is laid down along the center of the joist as shown. This type of fastening is only used with plywood sub-floors.

steel I-beam. Built-up girders are made up of two, three or four pieces of two-inch lumber nailed together with 20-penny nails. The joints, if any, should be staggered and only over columns or posts. Sizes of girders depend upon the load the floor will carry. Posts, whether steel or wood, should be equal in cross-section to the width of the girder they will support. For example, a 6x6-inch post should be used to support a girder that is six inches wide; if the girder is eight inches wide, then an 8x8-inch post should be used. Adequate footing must be provided for the posts or columns. A wooden post should be supported on a concrete footing which extends above the floor level (see drawing). For added security when installing wooden posts, a bolt, steel rod or pin, is usually installed in the footing before the concrete has set. This extends about three inches into the base of the post and guards against the post slipping off its footing. You can get a commercially made wood post anchor and support that will hold the post a couple of inches off the floor in order to protect the bottom from dampness.

One type of steel post has a heavy duty threaded stem that can be moved up and down to assure good contact with the girder above it. If the wooden structure above it shrinks, the post can be adjusted to take up the slack. *Caution:* Never take up more than a half-turn a week in order to permit the girder to reach equilibrium. Some local codes insist that the threaded portion of the post be made immovable by welding in order to forestall tampering by inexperienced owners.

When floor joists are set on a steel girder, a two-inch wood pad is often placed on the girder to carry the joists level with the sill plate. When a wood girder is used, it is usually set so that the top will be level with the sill.

SILLS

The sills, usually 2x6s, are secured to the foundation of the house with ½-inch or larger anchor bolts, spaced according to the local building code. Termite shields are a must if the house is in a termite area. The sill should be at least ten inches above grade; in quality construction, a sill sealer is placed between the sill and the foundation to take up any irregularities in the top of the foundation. A good idea is to drill the holes for the anchor bolts somewhat oversize so that any errors in measurement or misalignment will be "absorbed" by the oversize holes. Any low spots in the foundation should be brought up to level with additional concrete or mortar, though wooden wedges can be used if the low spots are slight.

The **floor joists** of a house transfer the load from the floor to the girders and the sill. They are usually two inches and sometimes three inches thick and are always placed on edge, spanning the distance from the sills to the girder. Joists must not only be strong enough to carry the load above them, but they must also be stiff enough so as not to deflect under load. The live load on a floor is generally figured at 40 pounds per square foot and most building codes insist that the deflection of a joist should not exceed 1/360th of the span with all appliances, furniture, people, etc. on the floor above it "in use." For joists spanning 15 feet at 16-inch intervals, this means that deflection is limited to ½-inch. Good construction calls for joists to be doubled around openings in the floor, such as for stairways, fireplaces and chimneys, and under room partitions.

Bridging is required by most local building codes, usually at 8-foot intervals. Bridging consists of small pieces of lumber set in a diagonal pattern to form an X as shown in the drawing. Bridging serves to keep the joists in a true vertical position and to transfer the load from one joist to the next. Bridging is installed during construction by nailing the pieces to the tops of the joists, leaving the lower ends hanging loose. These are not nailed until the subflooring has been installed.

Solid bridging is sometimes used in lieu of cross-type bridging, especially if the space between the joists is not 16 inches on center. Also called block bridging, this gives additional ri-

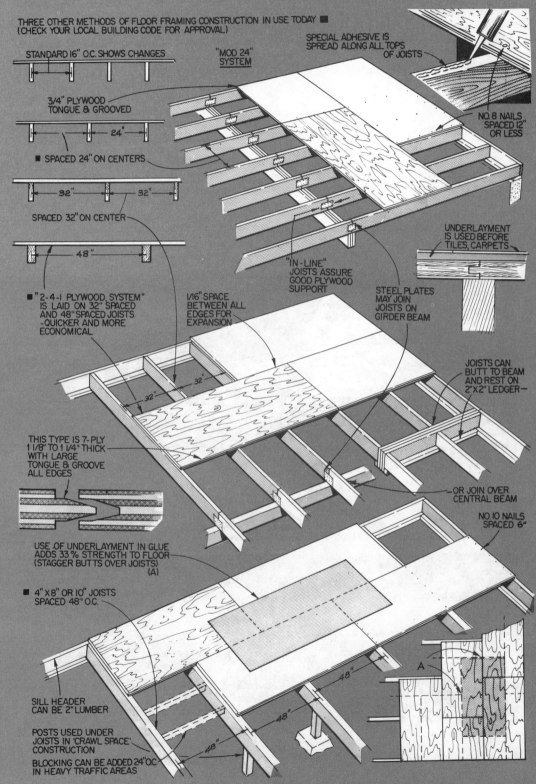

gidity to the entire assembly. Another type of bridging consists of sheet steel channels with a V cross section having teeth at each end. They are simply hammered in place—no nails are required—and meet FHA Minimum Property Standards as well as the Uniform Building Code.

SUBFLOORS

The final step in floor framing is the laying of the subfloor. This can consist of plywood, tongue-and-groove boards, shiplap flooring or even common boards just butted together. The subfloor adds considerable rigidity to the framing below as well as providing the base for the finished flooring (though sometimes "finished" flooring consists only of plywood when the floor will be carpeted wall to wall). The subfloor is also a working surface on which additional framing and construction is laid out.

Board subfloors are generally laid at an angle to the joists, starting about five feet from a corner and then filling in this corner with scraps cut from the longer lengths. The subfloor should be nailed at each joist with eight-penny nails. The ends should be permitted to extend beyond the walls and openings and then cut off after nailing has been completed. If plywood is used for subflooring, it should be laid so that the long dimension (the eight-foot side) is at right angles to the joists. Most plywood subfloors are 5/8-inch even though ½-inch plywood is approved by the FHA. Use eight-penny nails every six inches for installation; joints should always be over the center of a joist.

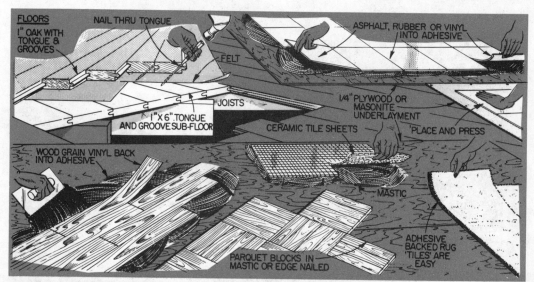

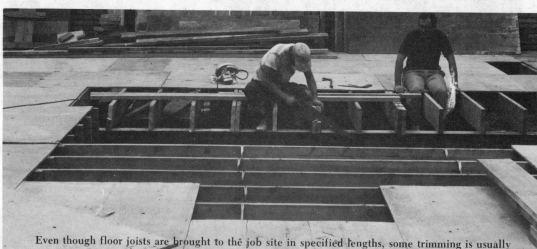

Even though floor joists are brought to the job site in specified lengths, some trimming is usually necessary. In an awkward position such as this, it is best to use a hand saw rather than power saw.

Note how this girder is supported by a steel lally column. Long girder spans should be supported every 10-ft. with such lally columns.

Left: applying sheathing to the 2x4 studding. Note how studs are reinforced with horizontal "cats" about halfway up from the floor.

WALL AND CEILING FRAMING

The second major area of house framing consists of the walls and the ceilings. The framing members are sole plates, top plates, studs, and headers or lintels. Studs and plates are usually 2x4s; headers may be doubled 2x4s or heavier timbers. When outside sheathing does not provide sufficient bracing, 1x4s are used in a diagonal pattern, recessed into the studs to provide extra rigidity (see drawing). Extra studs are always used at window and door openings, at all corners and where a partition meets an outside wall. Studs are normally spaced on 16-inch intervals (center to center).

The drawing shows two methods of corner construction. Note that in method A an extra stud is used to make the corner more rigid while in method B only three studs are used. Either method is acceptable. Many carpenters build a corner section independently of the wall studs. This way they can plumb the corner section to assure perfectly perpendicular walls.

When a wall meets a partition, or a proposed partition, extra studs are always used to provide backing for lath or wallboard (see drawing).

Headers support the weight of framing and ceiling over door and window openings. A header is made by nailing two pieces of lumber together, sometimes with a spacer in between to make the assembly equal the thickness of the wall. If the load above the opening will be especially heavy, a truss-type header may be used. Two common types of truss headers are shown in the drawing.

After erection of the wall framing, sheathing is installed. Plywood or fiberboard is generally used for this purpose. Both come in 4x8-foot sheets, though longer sheets are available if desired. Sheathing can be applied vertically or horizontally, fastened with 6-penny nails at six-inch intervals. Tongue-and-groove boards can also be used for sheathing. They are usually applied in a diagonal pattern, in opposite directions from each corner.

Ceiling framing is very similar to floor framing. The chief difference is that lighter lumber is used, unless the ceiling will also be the floor for the rooms above. Joist size is determined by the span they are to cover and by the load they will bear. Ceiling joists, like floor joists, usually run across the short side of the house. Where they extend to the edge of the structure and meet the roof, they must be cut to match the slope of the roof. This can be done after installation is complete and before roof framing begins. Where an access opening for the attic is required, doubled headers and joists are used.

Shown in the drawings are ten roof styles used in residential construction. Whatever the type of roof, it must support a heavy load—not only the weight of the sheathing and shingles, but also the tremendous weight of snow in northern areas.

There are several terms used in roof framing with which you should be familiar. *Common rafters* run at right angles to the wall plate. *Hip rafters* are set at an angle to both ridge and wall plate. *Valley rafters* extend at an angle from ridge to plate in the valley formed by the intersection of two dissimilar roof sections. *Jack rafters* are fully defined as *hip jack* (ends at a hip instead of the ridge rafter), *valley jack* (intersects a valley rafter instead of going as far as the plate), and *cripple jack* (a short rafter between hip and valley rafters). All are illustrated and their positions shown in the drawing.

Laying out a roof frame is a problem in geometry, but instead of working with pencil, ruler and paper, you work with lengths of wood to lay out combinations of right triangles. If you know the dimensions of two sides of a right triangle, you can figure out the length of the third side, by the simple formula that the square of the hypotenuse (the longest side) is equal to the sum of the squares of the other two sides. But you don't even have to do that, as a roofing square will give you the needed information.

The roofing square is an indispensable tool when framing a roof. Without the need for complicated mathematical formulas, it will determine the length of common, hip, valley and jack rafters for every pitch of roof. It will tell you how to make the top, bottom and side cuts for any rafter.

The square is made in the form of a right angle, with two arms; the longer one is 24 inches long and is called the body, the shorter and narrower one is called the tongue and is 16 inches long. Both sides of the square have a series of tables and scales mathematically worked out to give required information for making the various cuts needed in roof framing work.

Let's work out a simple problem using the square. The *run* of a common rafter is the shortest distance between a plumb line suspended from the center of the ridge to the outer edge of the roof plate. The *length* of the rafter required from the edge of the roof plate to the ridge is greater than the run, for it is the hypotenuse of a right triangle, which is always greater than either of its sides. The height between the ridge and the plate is called the *rise*. If the rise is, say, 10 feet, and the run is 15 feet, the square will

Applying roof sheathing. Butt joints should always be over a roof rafter, never between. Half-inch plywood can also be used for roof sheathing. A good job calls for nailing each rafter, no skipping.

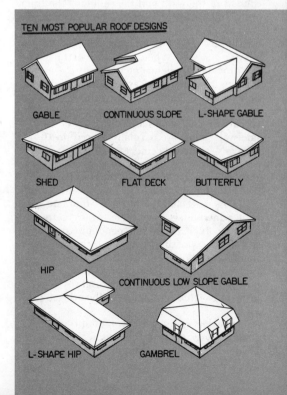

TEN MOST POPULAR ROOF DESIGNS

GABLE CONTINUOUS SLOPE L-SHAPE GABLE
SHED FLAT DECK BUTTERFLY
HIP CONTINUOUS LOW SLOPE GABLE
L-SHAPE HIP GAMBREL

quickly tell the length of the rafter needed to extend from the plate to the ridge. All you need do is place a ruler at the 10 mark on the square with the other end intersecting the 15 mark. Read the ruler and you will find that the length of the rafter required is 18, or 18 feet. If the rise is 8 feet instead of 10, and the run is the same 15 feet, then the length of the rafter will be 17 feet. What the square has done is to work out the mathematics for you. The square will also tell what angle to make the bottom and top cuts for the rafter so that it will meet the plate and ridge at the precise angle.

Openings must be made in the roof for chimneys and skylights. For such openings, double rafters and headers are always used.

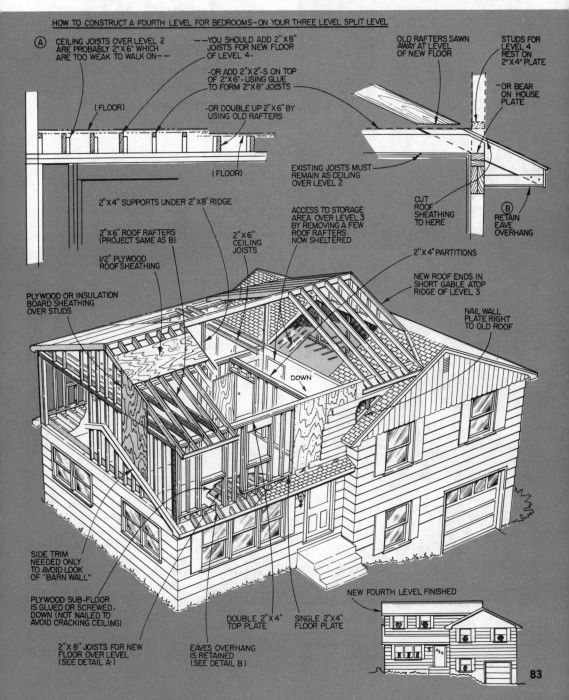

How to Hang Doors

How to cut new openings, reinforce framing, install hinges, latches and locks; tips and ideas; door problems and how to correct them.

An especially attractive entrance consisting of two, 24-in. pine doors. Though only the left-hand door is used for normal traffic, both doors can be opened to pass bulky furniture, large appliances.

The Dutch door, often used in kitchens, consists of two doors, each independently mounted. The lock and handle should be installed on lower section. Upper part is locked to the lower with latch.

ABOUT DOORS

If you are planning to install a door, you must know something about the kind of framing required. If you are going to cut through a load-bearing wall, such as an exterior wall, you should use a double 2x8 header above the door opening as shown on page 77. The header will carry the load formerly supported by the removed wall section. If the door will be installed in a non-load bearing wall, the header and the framework can consist of doubled 2x4s. Local building codes should be followed in either case.

But before picking up that hammer and chisel to cut the new door opening, make sure there are no heating ducts or pipes behind that wall. Electrical wiring is no problem as BX or Romex can easily be relocated (but be careful not to cut through it). Tap the wall or drill 1/8-inch pilot holes to locate the studs. If possible, have one side of the door close to an existing stud. The location of the stud on the opposite side will depend on the width of the proposed door. Cut the plaster or gypsum board with a chisel and hammer. If you come across wire lath, use a tin snip to cut it away; use a compass saw to cut wood laths. Next, cut away the center studs and the sole plate at the bottom. Removed studs can be utilized to make a double header.

Above are shown examples of doors suitable for outdoor use. Such doors are generally 1-¾ in. thick, are hung with three hinges and protected with a coat of varnish; are usually 36 in. in width.

Right: A door suitable for a house in Early American or Colonial styling. It is made out of white Ponderose pine and finished with an exterior grade of enamel. All windows are of stained glass.

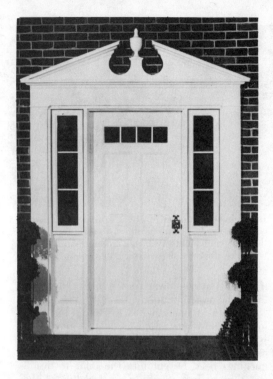

Doubled studs are used on each side of the door opening to minimize vibration from door openings and closings. The two studs on the inside of the opening help support the header. With the studs and header in place, the door opening should be 1¾ inches wider and 1¼ inches higher than the dimensions of the new door. Nail the head and side jambs in place, carefully checking with a level and shimming where necessary to make them perfectly plumb and level.

Because a door swings through an arc when mounted, the latch side of the door should be planed to a slight bevel toward the edge of the door that will meet the stop. The next step is to set the door in the opening using wedges at sides and bottom and allowing 1/16-inch clearance at top and each side. Use a chisel to mark the positions of the hinges, 7 inches from the top and 11 inches from the bottom. The chisel should be held so that it will mark the door and the door frame simultaneously. Next, remove the door and extend the hinge marks to the inside of the jamb with a try square and a pencil.

Draw the outline of the hinge on the door and on the jamb with a sharp pencil so the barrel of the hinge will extend slightly beyond the edge of the door. Fasten hinges to the door first. If you have a router, all you need do is extend the cutting length of the router bit to equal the thickness of the hinge leaf and rout away the wood so that each hinge will fit flush in its recess. If you do not own a router, score the outline of the hinge with a sharp chisel. Hold the chisel perfectly square and drive it down to the approximate depth (thickness) of the hinge leaf. Keep the beveled face of the chisel towards the cutout. When you have finished this outline, make a series of shallow cuts with the chisel starting at either end as shown in the drawing. Now hold the chisel along the depth line as shown and drive it lightly across the grain to remove the wood; the beveled side of the chisel should be up. Clean out the mortise, for that is what you have made with the chisel, and try the hinge for fit. It should fit flush within this opening. If it doesn't, keep up the good work. If too deep, it can be shimmed up with a bit of cardboard. Drill pilot holes for the hinge screws and attach the hinge with flathead screws.

By this time you undoubtedly will have noticed that the hinges can be taken apart by removing the hinge pin. The other half of each hinge is similarly fastened to the jamb. On solid, heavy doors and exterior doors, mount a third hinge midway between the other two.

Try the door for fit. It should swing freely without binding at any point. If it does bind, say at the upper corner opposite the hinges, insert a shim of cardboard under the lower hinge;

If you have a number of doors to hang, you can do as the pros do, use a hinge template. This time-saving device is used in combination with a router. It accurately indicates the exact placement of each hinge and then the router cuts away the exact amount of wood necessary to recess the hinge so that it will be flush with the door as shown in center photos. After mounting hinges, next step is to mark door for lock.

if it binds at the lower corner, insert the shim under the upper hinge.

The next operation is to install the door lock. The most commonly used lock nowadays is the standard **cylindrical lock**, sometimes called a key-in-knob lock. Only two round openings need be cut into the door to install this type of lock. Scribe a vertical line at the center of the door edge and a horizontal intersecting line at the required lock height, usually 38 inches from the floor. Continue the horizontal line to the front of the door with a try square, or use the template supplied with new locks. Next, using an expansive bit and a brace or a hole saw attachment in a power drill, bore the required hole opening in the door face (most locks of this type need a 2-1/8-inch hole). Drill the hole from both sides of the door to avoid splintering; wait until the tip of the bit starts to emerge from the opposite side of the door, then back out the bit and bore from the opposite side using the exit hole as the starting point.

The next step is to bore a hole in the edge of the door for the latch, usually 7/8-inch in diameter. Insert the latch in this hole and trace the outline of its faceplate with a sharp pencil. Use a chisel and a hammer to mortise (recess) this area. Next, install the strike plate on the jamb directly opposite the latch. The strike plate too must be mortised so that it will be flush with the jamb. More chisel and hammer work—unless you have a router. Fasten the strike plate with the screws provided.

Insert the latch into the door edge, making sure the tongue of the latch faces the right way, then slide the lock through the large hole in the door. Make sure the prongs of the lock properly engage the recess in the latch. Finally, finish the job by installing the holding plate and the knob. Try the handle to make sure the latch moves in and out freely.

Mortise locks take more time to install than key-in-knob locks, but they are more substantial and offer greater security. The body of a mortise lock is located within the door, so a recess must be cut into the edge of the door to house the lock. Determine the position of the lock on the door and bore a series of holes wide enough and deep enough to accommodate the particular lock you are planning to install. The holes should be about 1/8-inch apart; do not try to overlap them. Use a chisel and a hammer to remove the wood between the holes and along the sides of the resulting recess. You can use a *Surform* tool to smooth and enlarge the hole so that it will accept the lock without binding.

The next step is to slide the lock body into the recess. Mark the outline of the lock on the door edge and mortise for the top, bottom and sides of the lock front (another job made easier by a router). The recess should be just deep enough so that the front is flush with the surface of the door when it is secured with the screws supplied. Next, remove the lock body from its recess and mark the door for the required holes for the key cylinder and the door knob spindle. Replace the lock body and install the roses (round decorative trim plates) and trim for the handle and the key opening.

After the lock is in place, mark the door jamb for the strike plate and trace its outline. Mortise the jamb to accept the lock bolt and the spring-loaded latching bolt. (Better mortise

HOW TO CORRECT DOOR TROUBLES

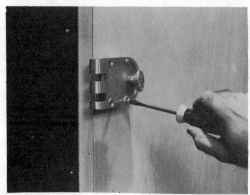

Probably the easiest lock to install is the night latch or rim lock as it is sometimes called. It only requires a single hole to be drilled through the door. Lock is adjustable for thickness of door.

A door may stick because of swelling wood, an out-of-plumb jamb, improperly installed hinges or the settling of the house. Open the door and inspect the hinges; all screws should be tight. If a screw or screws keep turning without tightening, remove the screw and insert a sliver of wood dipped in white glue—a kitchen match works fine, then re-insert the screw. Sometimes a longer screw will do the job of re-tightening. While you are at it, inspect the strike plate for looseness.

If the door still sticks, look for worn spots on the door jamb or the door. If the door binds near the top on the latch side, shim out the bottom hinge with a thin strip of cardboard. If the door is binding at the bottom on the latch side, check the upper hinge; it may have been mortised too deeply. Correct this by inserting a cardboard shim under the hinge leaf on the jamb. If there is binding along the top edge of the door, wedge it open and plane or sand down the top of the door. If the binding is at the bottom, you will have to remove the door and plane the bottom.

Suppose the door binds along the *entire* latch side. In this case, remove the door and plane down the *hinge* side. This way the lock need not be removed, and it will not disturb its relationship with the strike plate.

How about a spring latch that will not catch to keep the door closed? Correct this by shimming out the strike plate with one or two pieces of cardboard. Sometimes a door will open by itself when not latched, or will resist closing even though there is no binding. This trouble can usually be corrected by inserting a narrow cardboard shim under both hinges, but only on the pin side of the hinge.

locks have a dead bolt in addition to the spring-loaded latch bolt.) The dead bolt is moved in and out of the strike recess by means of a small handle or knob on the inside of the door. Close the door and see if the dead bolt and the latch work smoothly. Try the key. If necessary, the strike plate can be adjusted by filing the opening, or if it is greatly out of line by re-positioning it.

Rim locks or night latches as they are sometimes called, are the simplest of all locks to install. They are mounted on the surface of the door and only one hole through the door is required. Determine the lock height, usually about 50 inches from the floor, and drill a hole of the size indicated by the manufacturer of the lock. Distance from the edge of the door is also determined by the lock make. Install the lock cylinder on the outside of the door and the connecting back-up plate on the inside. Secure it with the bolts furnished. These bolts have annular rings at half-inch intervals to help you cut them off to match the thickness of the door. Screw the lock to the inside of the door with the wood screws provided. This completes the installation except for the mounting of the strike plate which is fastened to the door jamb. Close the door and mark the position of the strike plate on the jamb so that it is in line with the lock. Open the door, place the strike plate in the marked area and trace its outline with a sharp pencil. You will have to mortise the strike plate so that it is flush with the inside of the door jamb, as well as flush or partially recessed in the door trim. Finish the job by securing the strike plate with the wood screws provided. Try the key to make sure there is no binding as the bolt is extended and retracted.

DOOR TIPS

File a notch on the top of the hinge barrel before installation. If it ever becomes necessary to remove the door for repair or painting, a screwdriver placed in the notch will make it a cinch to drive out the hinge pin.

When planing a door, always plane to a reference mark to avoid taking off too much wood.

Installing Kitchen Cabinets

Whether you make your own or buy them ready-made, installation is fairly simple.

Custom-matching kitchen cabinets are available to match the trim on dishwashers and other built-in kitchen appliances. Trim and matching cabinets are also made in stainless steel, white, copper, and most of the other popular colors.

Kitchen cabinets are available in so many size combinations that it is a comparatively simple matter to fill a wall from end to end and not have a gap anywhere. Filler strips are available just in case you can't make a combination of cabinets "fill" a kitchen wall.

The first step in hanging cabinets is to install two parallel furring strips across the wall. These strips, one for the top of the cabinets and the other for the bottom of the cabinets, must be screwed or nailed to the wall studs. In new construction, the builder will generally nail these strips across the studs before plastering. He will then tell the plasterer or dry wall installer to mark the positions of the furring strips with a pencil or a scratch mark. This way the kitchen cabinet installer will not have to probe for the furring strips. If the walls are already plastered, there are three ways you can locate the studs. The first is by tapping the wall. You don't need an ear like Tòscanini had to be able to tell a "hollow" tap from a "solid" tap; the solid tap is where the stud is. The second method is to drill a series of 1/8-inch holes with an electric drill, an inch apart. You will be able to tell by the feel when you are drilling into a stud or when you are drilling into plaster and there is no stud behind it; the drill passes into empty space without any slowing. Incidentally, just because you have located a stud this way, this doesn't mean that you have located the nailing point. The drill may just have hit a scant 1/16-inch from the edge of the stud. No good—you want to nail into the approximate center of the stud. So, drill another hole, an inch away from the first. If this second hole also strikes the stud, you have located one stud, others should be at 16-in. intervals, but check by drilling at these locations. It is most important that the furring strips for supporting suspended cabinets be securely anchored to studs. The third method of locating a stud is recommended only in desperation, or if your plaster is in such poor shape that it must be replaced anyway. That is to tear away the plaster—and there behind it you will find the stud!

Now that you have located the studs, attach the two furring strips. The top strip should

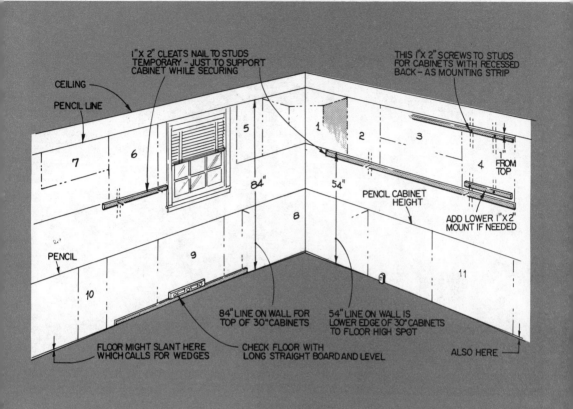

be about an inch below the top of the cabinets; the lower strip about an inch above the bottom. The strips should be cut so they are an inch scant at each end. Make sure the furring strips are straight, check with a level before pounding that second nail in place. This method can be used with all kitchen cabinets having recessed backs. The cabinets, when screwed to the furring strips, will then be flush with the walls.

Each manufacturer has his own theory as to how a cabinet should be built and so some cabinets have flush backs. In this case install the furring strips so they will be flush with the top and bottom of the cabinets. The furring strips will then be used as a base for attaching a finishing molding. Cabinets should be fastened to the furring strips with screws rather than nails.

Wall cabinets are almost never mounted flush to the ceiling (too high for easy access). You can leave the space above open to accumulate knick-knacks and dust or you can enclose it for extra storage space. Sliding doors will keep dust away from the stored items and present a tidy and pleasant appearance to the tops of the cabinets. Another possibility is to install a soffit to hide this area from view. As shown in the drawing, the soffit is flush with the front of the cabinets. Molding hides the joint between soffit and cabinets.

Base cabinets are installed similarly to the wall-hung types, unless they are free-standing, in which case no furring strips are required. Quite often allowance must be made behind base cabinets for electrical outlets, for a gas line to the range or for water pipes to the dishwasher. If such is the case, you can use doubled furring strips, or even 2x4s. But the final installation of the base cabinets should place them in line with the kitchen appliances and fixtures—the sink, washer, range, etc.

Drill a series of holes, six inches apart at the top and bottom of the cabinet to match the spacing of the upper and lower furring strips. The holes should be large enough to pass a No. 8 screw without binding. You can use flathead or roundhead screws; if you use roundhead screws, place a washer under each screw; if you

The kitchen is your wife's workshop; make it pleasant for her, and it will be reflected in the meals she prepares for you. After appliances, cabinets are the most important adjunct to an orderly kitchen.

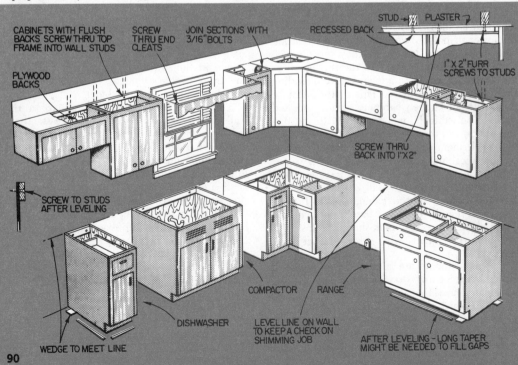

In order to avoid tedious and painstaking trimming, this Armstrong Solarian floor covering was installed before the floor cabinets were placed. Next step, shim the cabinets to make them level.

After all cabinets have been installed and aligned, bolt them together, even though they may be free-standing, and replace all doors, if they have been removed. Next, step back and admire!

opt for flathead screws countersink the opening so the screw will be flush with the inside of the cabinet when driven home.

Place a level at the bottom of the cabinet (doors should be open, or removed if you prefer) and have a helper hold the cabinet at what you think is the right height. Get in a position where you can see the top of the furring strip, or a mark placed nearby to indicate its position. Now drive the first screw at one end, but not too tight. Have your helper consult the level and move the free end of the cabinet up or down as the level indicates. When the level shows the bubble in the middle, tell him to hold it while you drive in the second screw at the opposite side of the cabinet. Then drive the rest of the screws in place, top and bottom. Make a further check with the level to be sure that the cabinet has not been knocked out of line. Go over all the screws to make certain that they have all been driven home securely.

After all the cabinets are in place, bolt them together through their sides. This is necessary to avoid spaces between the cabinets possibly appearing due to settling of the house or an extra heavy load in an adjoining cabinet. Drill holes at the front edge of each cabinet, top and bottom, and insert 3/16-inch bolts with washers.

Wood cabinets come with or without bases. If you have to make a base for a cabinet it can consist of a simple framework of 2x4s as shown in the drawing. The floor must be level, or the base will have to be shimmed with wedges until the bubble in the level is in the exact center. Install the floor cabinet on its base (if a base is required) and push it against the wall. Start at one corner of the kitchen, working toward an appliance such as the range or dishwasher. After all the floor cabinets have been properly leveled, drive screws through the back into the furring strips. This installation is somewhat easier than wall cabinet installation as you can do the work without help. Join the cabinets together at their forward inside edges with short bolts, washers and nuts.

Steel cabinets. Steel cabinets generally have some sort of hanging bar in the rear for attachment to the wall. The hanging bar is installed at the proper height and then the cabinets are lifted up and lowered over the bar. In effect, they are hooked in place. Some steel cabinets have holes in the back for attachment directly to the studs with roundhead screws.

Counter top covered with Goodyear's flexible vinyl. It is made in 36-in. widths and in 18 colors. It comes in roll form and is especially suitable for the do-it-yourselfer; used with a special adhesive.

This butcher block is really a plastic laminate made by Formica to resemble wood. All of these counter tops require practically no maintenance. Caution: do not use in lieu of chopping block.

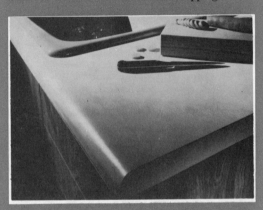

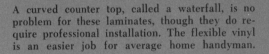

A curved counter top, called a waterfall, is no problem for these laminates, though they do require professional installation. The flexible vinyl is an easier job for average home handyman.

How to Cover Countertops

The traditional plastic laminates plus a newcomer, durable vinyl in roll form that is easy to install and stands up to hard use.

At one time most countertops were covered with a plastic laminate. Nowadays you can get vinyl in roll form, in 36-inch widths and in many patterns, especially made for do-it-yourself installations. The first step is to cut the base for the countertop out of ¾-inch *exterior* grade plywood—after all, countertops do get wet. Make sure that you get plywood with at least one good side, free of knots, cracks and similar flaws. The vinyl must be applied to the good side, otherwise imperfections will "telegraph" and be visible as shadow marks on the vinyl.

Store the vinyl for at least 24 hours in a room where the temperature is at least 70°.

Sand the good side of the plywood thoroughly. Now spread the special adhesive that is sold with the vinyl, using a notched trowel. Unroll the vinyl over the plywood, starting at the most convenient end. Cut off the surplus, allowing some excess. Lift up one side of the vinyl to make certain that it is getting a good transfer of adhesive from the plywood. If gaps have developed, you haven't applied enough adhesive. Spread some more on any dry spots, and roll the vinyl—a rolling pin is fine. Or use a dampened cloth, rubbing forcefully from the center outward to all edges. Any excess adhesive can be wiped away with a water-dampened cloth.

The main point in using plastic laminates in the kitchen, or other rooms for that matter, is the ease with which it is cleaned. Just a swipe with a damp cloth does the job; makes your home maintenance-free!

However, adhesive that has already dried can only be removed with a cloth dampened with mineral spirits, such as paint thinner.

After a couple of hours' drying time you can trim off the excess vinyl around the edges; use a sharp razor-thin knife and cut downward to avoid possible lifting of the vinyl.

The edge of the countertop can be trimmed with metal molding, wood molding, or with the vinyl itself. If you elect to trim with the vinyl, allow the countertop vinyl to project over the edge about ¼-inch. Cement the edge vinyl, butting it tightly against the top vinyl. The top vinyl should then be trimmed with a sharp knife—after a two-hour wait—so that it is flush with the vinyl on the edge.

If you have to seam the vinyl, as when making a U-shaped or L-shaped counter, make a double cut seam (also called a Dutch cut). Apply adhesive to all but about four inches from the meeting edges. Overlap the vinyl about three inches and cut through both sections with a sharp knife, using a steel straightedge as a guide. You may have to make several repeat cuts to get through. Persevere! Now remove the cut-away material, apply adhesive right up to the edges and press down firmly using a roller. Place a weight over the seam until the cement sets.

To make a vinyl-covered countertop for a sink, turn off the water at the main, disconnect the drain and the faucets and remove the sink from the counter after detaching the pipe lines. Cover the entire counter area including the opening for the sink with the vinyl, as described before. Best to wait until the next day to make certain the vinyl is good and tight to the counter before cutting the opening for the sink. Use a sharp knife. Start at the center by making an X cut, then draw the knife to each side of the sink opening, removing a series of pie-shaped pieces.

Now apply a generous amount of a waterproof calking compound to the underside of the sink lip. Insert the sink in the opening and wipe away any excess compound.

Facts About Fasteners

Nails come with heads, without heads, double heads, serrated, square, with annular rings and in copper, steel and galvanized.

The trick is to select the correct one needed to perform the necessary task. Get to know what they are, what they will do.

It is for good reason that the nail is sometimes called The Great Joiner. There are literally hundreds of nail sizes and types made for home and industry. The length of a nail is designated by inches and by "penny" size. This term originally referred to the price per hundred; the larger the nail the more pennies per hundred, but now nails are sold by weight and penny signifies only length. The chart shows the relationship of penny size to inch size.

NAILS

The diameter of a nail increases with its length, except for special purpose nails such as floor nails and shingle nails. In addition, nails are distinguished by their heads—a large flat head such as on the common nail and a small head, slightly wider than the shank diameter, on a finishing nail. The large heads hold best because the load is distributed over a larger area but the small heads are easier to conceal by driving them slightly below the surface of the wood with a nail set and covering the holes with putty. Another advantage of small-head brads or finishing nails is that they can be drawn through the work from the inside without marring the surface when it is necessary to disassemble a cabinet, for example.

As a general rule when joining two pieces of wood, nails should be driven through the thinner piece and into the thicker piece. For maximum holding power, drive the nails at a slight angle. If you can "clinch" the nails, so much the better, but you will need nails long enough to pass completely through the work with at least ½-inch of nail protruding.

The chart shows some of the many nails and other fastening devices applied by hammering and screwing.

SCREWS

Screws are more sophisticated fasteners for joining wood and parts to wood. Screws are used where greater holding power is required and also when disassembly is a factor. The most common type of screw has a single slot across its head. The Phillips-type screw has a cross-shaped slot and requires a special screwdriver. Phillips screws reduce the danger of the screwdriver slipping from the recess and possibly marring the work.

Screws are sold by length, diameter and head type, as well as by type of metal (brass, steel, aluminum, stainless, bronze) and plating (cadmium, zinc, chromium or other metal). The length of a screw is measured in inches (until we adopt the metric system) while the shank diameter is indicated by a number, ranging from 0 to 24. The greater the number, the thicker the diameter of the screw (see chart). The screws most commonly used around the house range

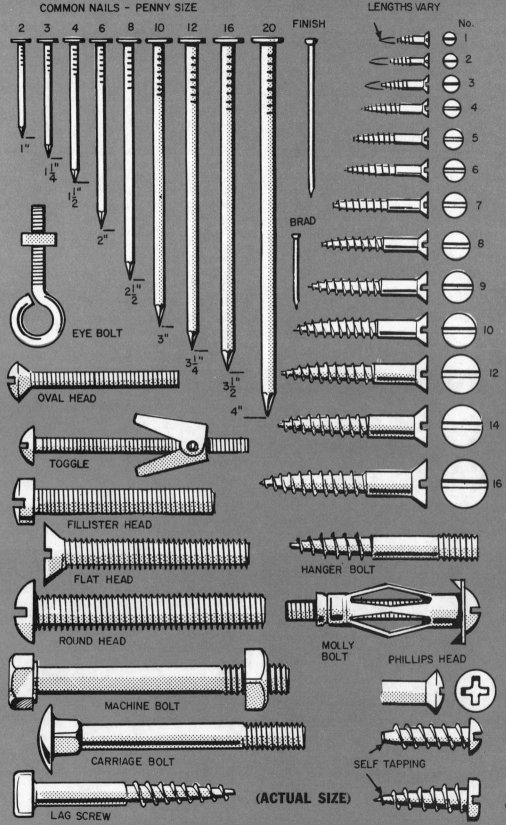

from Nos. 6 to 10. The three most common heads on screws are the oval head, the flathead and the roundhead. Roundhead and oval head screws are used when the screw head will be exposed, while flatheads are used where some concealment is required.

Small screws can be started by making a pilot hole with an awl. Larger screws should always have a pre-drilled pilot hole. Pilot hole size is shown on the chart. When using large screws, and especially when working with hardwood, a clearance hole for the shank will also be required. Use the chart as a guide. To avoid repetitious bit changing in your drill, you can get a set of bits called *Screw-Mates* which will drill a pilot hole, a clearance hole and even a countersink for a flathead screw, all in one operation. This is a great time-saver when a number of screws of the same size must be driven.

When driving screws into hardwood, lubricate the screw threads with wax (soap tends to rust the threads). Washers are sometimes used under round and oval screw heads to provide extra bearing surface and to prevent marring the work when removal is required for access to inner parts, such as in a hi-fi cabinet. Match the washer to the screw head as shown in the drawing.

Hardwood "plugs" are sometimes used to conceal screw heads as, for example, in a "pegged" floor. The technique is to counterbore for the screw head—about ½-inch is sufficient—and then to tap the plug into the hole. Plugs are easily cut—sometimes out of matching wood or even out of contrasting wood—with an inexpensive device called a plug cutter mounted in a drill press. Plugs can be removed by drilling out the center and then collapsing the sides with a narrow chisel.

Blind nailing is a technique not too well known despite its simple operation. First step is to lift up a sliver of wood at the point where the nail is to be driven. Cut outline with sharp knife.

If possible, leave the sliver of wood hanging by its end, like a hinge. Drive the nail in place, use a brad or finishing nail. Sliver, of course, should be wide enough to completely cover nail head.

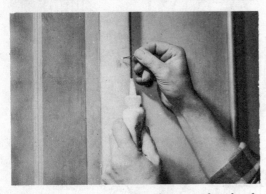

Next, apply a dab of white glue to underside of the sliver. You don't need much glue. After all, you are not gluing a couple of 2x4s together. Little bit of glue applied with another nail will do.

We are approaching the home stretch! Apply some masking tape over the sliver. Pull it tightly to hold the sliver snug against the parent wood. Allow the glue to dry before removing the tape.

This tool is used to drive nails into masonry. It holds the nail perpendicular to the wall surface and at the same time prevents it from buckling. Nails are available for it in many lengths, also threaded bolts.

This device is used to hold a corrugated fastener in place while it is being hammered home. These fasteners serve to reinforce miter joints when making wooden frames for storm windows and screens.

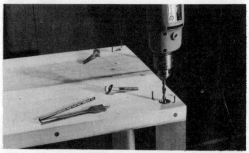

Drilling a hole for a lag bolt recess. When a bolt or a screw is to be driven into end grain, an extra long bolt or screw should be used. Note the recess at the top which will hide the bolt head from view.

Socket wrench drives the lag bolt home so that the head will be below the surface of the work. After head is driven below surface, a wood plug is inserted to hide the bolt and dress up the table. Neat, huh?

MORE FASTENING DEVICES

There are many other devices for fastening things together: bolts, lag bolts, *Molly* bolts, jack nuts, *Chevrons*, corrugated fasteners, U bolts, J bolts, eye bolts, turnbuckles and toggle bolts, to name the most popular.

Bolts, the most common fastening device after nails and screws, are designated by diameter in inches and by the number of threads per inch. For example, a ¼x20 bolt is 1/4-inch in diameter and has 20 threads per inch. Its third dimension is its length in inches. Bolt sizes range from 1/8x40 to ½x13 and from ¼-inch to six inches in length. Larger sizes used in construction and industry are seldom stocked by hardware stores. When using a bolt with wood, always place a washer under the nut to prevent its digging into the wood. Use the chart to find the fastening device best suited to a particular problem.

Sanding Techniques

Your workmanship is usually judged by its final appearance. Sanding is the last step before finishing and the most important.

A belt sander makes short shrift out of sanding large areas, such as this flush door. If you do decide to go first class, buy one with a vac attachment to suck up the dust; worth the extra cost.

It's the last sanding that determines the final finish. In order to do a good sanding job, you must know something about sandpaper. Actually the word "sandpaper" is a catchall for the many types of abrasives mounted on paper and cloth. The grades of sandpaper range from very coarse to very fine. The old grading system uses grit symbols. The coarsest is called 4¼ and the finest is labeled, 10/0. The more modern system uses numbers which represent the openings per inch in a screen through which the abrasive grains can pass. These numbers range from No. 12 (very coarse) to No. 600 (very fine).

Modern coated abrasives include silicon carbide (the hardest and sharpest), aluminum oxide, garnet, flint and emery. They may be mounted on paper or on cloth; may have the grains close together or far apart; may be used dry or while wet. The least expensive sandpaper is flint, a natural quartz material light tan in color. It clogs quickly, but because of its low cost can be discarded after a few minutes of use. This is a good paper to use for cutting down painted surfaces or pitch-loaded boards. Clean, fresh wood is best sanded with garnet paper. Thus too is an inexpensive material. When power sanding, use aluminum oxide paper. It costs more than garnet or flint, but lasts longer and is less expensive in the long run. Silicon carbide is used on metals and plastics; for polishing metal, emery is the usual choice.

There are many grades and types of sandpaper available, but if you stick to the five described below, they will take care of 99 per cent of your needs.

Very fine. Use this grade for sanding between coats of paint, varnish and lacquer. It yields an extra smooth finish and can be used wet for metal and dry for other surfaces.

Fine. Use this grade for final sanding before the first coat of a primer or sealer; also use on metal to remove light rust and imperfections.

Medium. Use this for light stock removal by

The same sanding disc can also be used to sand a slightly concave surface, such as this chair seat inasmuch as the rubber disc is somewhat flexible. Finish by hand using extra-fine paper.

A sanding disk chucked into a ¼-in. drill can be used to remove blistered paint from siding prior to painting. Follow up by hand using a medium-grade sandpaper to "feather" sanded areas, then paint.

For fine sanding before staining and finishing, wrap a piece of extra-fine sandpaper around a block of wood and rub with the grain until entire area feels smooth, remove dust before staining.

When it comes to refinishing a floor, nothing can beat a commercial floor sanding machine which you can rent by the day. A dusty, dirty job; remove all furniture and drapes before starting.

power or hand sanding; also on walls prior to painting or papering and for removing rust stains on metal.

Coarse. For rough stock removal, chiefly used with power belt sanders; it will smooth out deep scratches and similar imperfections before medium grade sanding.

Extra coarse. Used for removing heavy coats of paint and varnish, especially from floors; nearly always used with a power sander, belt or disc; also suitable for removing heavy rust deposits. Extra coarse sanding must always be followed by finer grades of paper to smooth the scars made by this grade.

All sanding should be parallel to the grain of the wood. Before starting to sand with the next finer grade of paper, wipe away the dust left by the coarser grade. And before starting to paint, varnish, shellac or lacquer, wipe the work thoroughly with a rag dampened with the solvent of the material you are going to apply. This serves to remove all dust and also gives the finish a better "bite."

A finished ceiling using the Integrid system. Unlike conventional tiles, this ceiling has no edge bevel to interrupt the texture. All supporting metal is concealed; the job looks like a one-piece installation.

High Points on Ceilings

There are almost as many ceiling treatments as there are wall techniques. Avoid conflict, choose the one your wife likes best!

MOST CEILINGS are simply in need of paint. Some minor repairs may be necessary, such as tapping in nail heads, spackling them as well as those hairline cracks along the molding trim. Spackled areas should, of course, be sanded. A flat white acrylic paint applied with a roller should do the job in one coat. Long-handled rollers are available, which eliminates the need for step ladders, but these are sometimes difficult to use due to the added stress needed to work them as well as having your eyes some distance from your work. You could miss a spot or not cover as well as if you were right up there.

Badly cracked or patched ceilings can be improved with a coat of sand paint. This is made especially for the purpose. The finished product is a textured surface which hides all flaws. The same job can be done with ceiling paper, with interesting effects. Of course, if paint or paper is not the solution to your ceiling problem or if you have an unfinished area that requires a completely new ceiling, then you should go the tile route.

INSTALLING A SUSPENDED CEILING

A new system, called *Integrid*, marks an important breakthrough in residential acoustical ceilings. Not only is it easy to install, but the finished ceiling is handsome enough for use in the "best rooms" in the house. What's more,

1. First step is to nail up the metal furring channels, 26" from wall; others are 24" on centers.

2. Channels are designed to accept the cross tees, which in turn hold the tiles. Tees clip to channel.

3. After all channels are in place, install first tile; next, snap a 4-ft. cross tee onto channel.

4. Continue across the room installing cross tees and tiles. Save leftover pieces to start next row.

5. This system can also be used to install 12"x12" tiles. Procedure is same, regardless of the size.

Integrid system can also be used for suspended ceilings. Start by nailing molding to all four sides of the room at the height you want the finished ceiling to be. Metal or wood molding can be used.

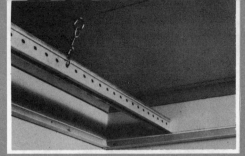

The next step is to install the main runners on the hanger wires as shown. The first runner is placed 26" from the wall; others are spaced 48" apart and perpendicular to the direction of joists.

the system requires no complicated room layout.

Integrid differs from conventional suspended ceilings in that the metal suspension members are "integrated" right in the ceiling itself. Specially fabricated 1 x 4-ft tiles butt tightly together to form a continuous, unbroken surface. Because each tile covers four square feet, there are fewer tiles to handle, which means easier, faster installation.

There's also a unique economy factor built into the system. When a ceiling tile is cut to finish out a row, the leftover piece, instead of being discarded, is used to start the next row. This eliminates tile waste.

Three different installation methods are offered with the *Integrid* system, each designed to meet a specific job need:

Full Suspension System: For rooms where a dropped ceiling is desired, *Integrid* provides the convenience of a full suspension system. Simply nail wall molding to all four walls at the desired ceiling height, and hang the main runners. Then, starting in a corner of the room, lay the first tile on the molding clip a four-foot cross tee onto the main runner, and slide it into a special slot on the edge of the tile. Continue across the room in this manner inserting tiles and cross tees. Since the tiles have no edge bevels, and since all supporting gridwork is concealed, the finished ceiling comes out looking like a one-piece installation.

Direct Integrid: There's also an optional direct-attachment method available with *In-*

CEMENTING ACOUSTICAL TILES DIRECTLY TO OLD CEILING

If the old ceiling is still relatively smooth though somewhat unsightly, new tiles can be cemented directly to the plasterboard. This type of installation is fast and easy.

Plasterboard, sometimes called dry wall or Sheetrock, can develop several unfortunate maladies as a ceiling material. Seams can separate and become visible, and nail heads pop into view. Leaks may cause dark water spots to appear.

On the other hand, tiling over a plasterboard ceiling is a task that any reasonably handy man, or woman, for that matter, can do with a minimum of tools and supplies.

The ceiling of a 12 by 15-ft. room can be covered with deluxe *Chandelier* acoustical tiles for as little as $65, including adhesive. *Chandelier* tiles are relatively new and fit together so closely that the ceiling looks like a one-piece unit, rather than a series of individual squares.

To install, daub ceiling cement on the four

As these pictures show, even the lady of the house can apply ceiling tiles—with her husband's encouragement of course. First step is to apply a dab of the special cement recommended to each corner.

After all runners are in place, start installing the tiles in a corner of the room. Just lay the first tile on the molding. Next, snap a 4-ft. cross tee onto main runner; tee fits into concealed slot.

Finished ceiling presents a one-piece look. This system is ideal for hiding old, cracked ceilings and the fact that tiles can be removed for wiring or replacing pipes makes it ideal for basement.

tegrid, designed for use under exposed wood joists. In this type of installation, the 1 x 4-ft. tiles attach directly to the joists by means of a newly developed Armstrong joist clip. No hanger wires, furring strips, or main runners are involved. Instead, a special spring-clip is inserted over the bottom of a joist, and the supporting cross tee merely snaps onto the clip. Only one clip and one cross tee is needed to support four square feet of ceiling tile.

Remodeling Installations: Integrid can even be attached to an existing suspended ceiling, affording unprecedented cost savings in remodeling and renovation work. Instead of removing the old grid system and panels, the installer merely attaches the *Integrid* cross tees to the existing ceiling main runners. Once these are in position, the new tiles slip into place exactly as they do with suspended and the *Direct Integrid*.

Accessibility to areas above the ceiling is achieved by fabricating special access panels where needed. This is done by removing the tongue-and-groove joint on the access tile, then screwing the tile to slots in the supporting cross tee. To remove the panel, simply unscrew it from the tee and lift it out of the ceiling.

A special recessed fluorescent light fixture is also offered with the new *Integrid* system. Available in a choice of two- and four-lamp models, the unit mounts easily on the *Integrid* main runners. The system can also accommodate conventional hanging fixtures.

Next, slide the tile into place so that the lip of the tile fits under the ledge of the tile already installed. Be careful here, make sure that the pattern of the tiles follows proper sequence.

Last step, just for added insurance to make sure the tiles stay in place while the cement sets, drive a couple of staples through exposed flange. Your husband should be proud of this job when finished.

corners and center of a tile. Press the tile into position on the ceiling. The tiles have tongue-and-groove joints to facilitate assembly.

To hold a tile in place while the cement "sets up", put two staples in each exposed flange using 9/16-in. staples.

The finished ceiling is not only pleasing to look at but adds comfort to a room from the standpoint of sound conditioning. Remember, acoustical tiles will absorb more than 50 per cent and some as much as 75 per cent of the noise striking their surface.

Storing clothes, blankets, and even socks, are the only way to keep them from becoming a meal ticket for ravenous moths and insects. Keep clothes lightly spaced so that cedar aroma will permeate the garments fully.

Build This Cedar Closet

This is a very special project which will please every member of the household. It can be free-standing or built in.

This cedar closet was built between two walls of a basement play room. It can just as well be a free-standing unit or even a room divider.

Red cedar for lining closets comes in strips two to four inches wide, 3/8-inch thick and up to eight feet long. These strips come with a tongue and a groove to make a tight, interlocking construction.

Make a foundation for the closet of conventional 2x4s, nailed together on edge as shown in the drawing. Next install the floor, the sides, the back and the ceiling. Use exterior grade ¾-inch plywood for a basement installation, unless you are sure your basement is reasonably dry. If the closet is well above grade, in the dry part of the house, you can safely use interior grade plywood. The two doors are inexpensive hollow-core doors mounted on 2x4s as shown. If you prefer, you can make doors of plywood, plywood will cost nearly as much as hollow doors.

After the shell of the closet has been completed, line the closet—and the doors—with the red cedar strips. Use one-inch brads to secure the cedar lining. Apply the strips horizontally as shown, in random lengths. You can install them vertically if you prefer, but the easiest way is to start at the bottom of the closet and work upwards. Cover the floor and the ceiling with the cedar strips as well as the walls. Since hollow-core doors have a rather thin veneer, it is advisable to use either a cartridge-type cement or a floor tile cement, as well as the brads, when installing cedar strips on the doors.

Install a cedar shelf and a clothes pole as shown. Make certain when you install the pole, that garments, when hung, will not rub against the rear wall. Add a couple of magnetic catches to keep the doors closed. For a finishing touch, you can add molding around the exposed edges of the closet. The outside of the closet can be either painted or stained and varnished—*but do not finish the interior.* To make the closet absolutely airtight, and keep the aroma of the cedar "locked in," add vinyl or felt weatherstripping around the door openings. After a few years you may notice that the aroma of the cedar is not as apparent as it was when you first built the closet. You can restore it to its original freshness by lightly sanding the surface of the cedar.

And incidentally, it is always a good idea to dry clean clothes, or at least give them a good airing in the sun, before putting them away for storage in the closet.

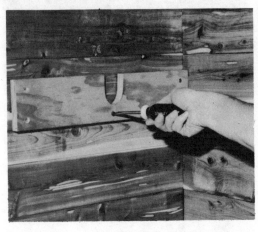

The cedar strips are nailed to the previously installed framing with 1" brads. The strips have a factory-cut tongue and groove facilitating installation. Set brads below surface so they won't catch clothes.

The clothes pole holder is cut from a piece of scrap cedar. Saw a semi-circular opening for the pole and then screw it to the wall with No. 8 flathead screws. Sand lightly to remove any splinters.

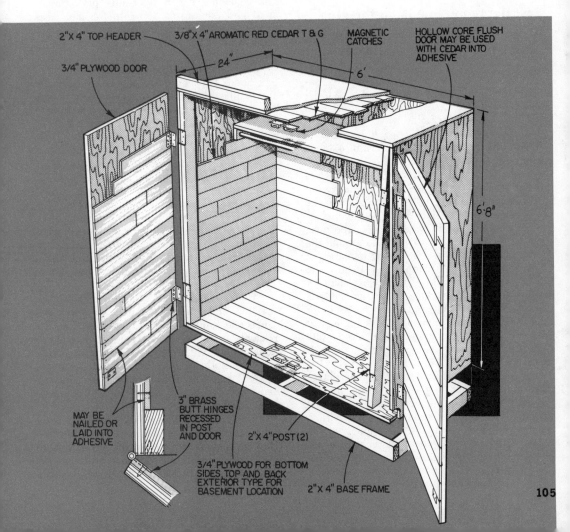

How you can use a C-clamp to apply pressure to a glue block used to reinforce furniture corners. Make the right-angle out of ¾-in. wood and cut off the corner to make a flat surface for the C-clamp.

Advice on Adhesives

List the requirements of the glue to be used. Select the right one and apply it as recommended.

Quite often it is undesirable, or even impossible, to fasten things together with such devices as nails, bolts and screws. You must then turn to glues and other adhesives. First consider where the item to be glued will be used or placed. For example, outdoor furniture joints require waterproof adhesives, not water-resistant ones. Generally, water-resistant adhesives are satisfactory when used in kitchens and bathrooms. Following is a description of the more commonly used adhesives and cements.

White glue, or PVA adhesive, is the most popular for the home workshop and around the house. These adhesives dry clear, but are not waterproof. Excessive dampness will cause the glued joint to turn milky. They are excellent for most interior woodworking joints and general household repairs. They can also be used for gluing paper, leather and cloth. Excess glue can be cleaned off with a damp cloth. Use moderate pressure when clamping joints; the glue sets, ready for use, within four hours.

Epoxies are especially suitable for gluing hard materials such as glass, china and metals. Usually they are formulated so that an equal quantity of resin is thoroughly mixed with a hardener and the resultant mix applied to the work. Epoxies cure by means of a chemical reaction between the "mixes." They are good for general repairs, but rather expensive, especially where a considerable quantity is required. Some epoxies dry clear, some white and some gray; they are available in tubes, cans, and even in gallon containers for industrial use. Curing time is generally 24 hours, though fast-curing types are also made.

Contact cement is used to bond plastic laminates to countertops, walls and other surfaces. It can also be used to bond metal, hardboard and plastics to wood. It should not be used over a painted surface, as it will cause the paint to lift, weakening the bond. It is not recommended for furniture joints.

Contact cement is applied to both surfaces to be joined and allowed to dry for about 15 minutes; the pieces are then pressed together. The work must be accurately positioned—once contact is made, it is very difficult to move the pieces. A paper "slip sheet" is usually placed between the pieces while lining them up. Excess cement can be removed with nail polish remover or special solvent. It is possible to separate cemented panels by gently prying them apart and pouring solvent into the gap; keep prying, using a wedge if necessary, and applying solvent. The operation is time-consuming, but it works.

Rubber-base cement makes good to excellent bonds between wood and concrete, paper and wallboard, and for pottery repairs. It is a "thick" cement, good to use when a filler is required between the mating pieces. Clean up with nail polish remover or acetone.

Plastic cements are for general household use. They are highly resistant to water. Work should be clamped while the glue hardens. Clean up with acetone before the cement dries.

Latex-base adhesives are used chiefly for gluing fabrics, carpets, cardboard and paper. They form a strong flexible bond that will withstand washing but not dry cleaning. Excess adhesive can be cleaned off while still wet with a water-dampened cloth; when dry, use lighter fluid or paint thinner.

Heavy-bodied **mastic adhesives** are used to bond ceiling tiles, floor tiles, plywood wall panels and similar building materials. They are available in tubes, gallon cans and five-gallon containers. There are two types: latex and rubber resin. Both are used for cementing to wood, painted walls, hardboard, tiles and concrete. Remove excess with a damp cloth; if dry, with paint thinner.

Gums and pastes include rubber cement, flour pastes, animal glues and vegetable glues. They are chiefly suitable for paper, cardboard and leather work. Most wallpaper adhesives fall into this category. Clean off excess with a damp cloth.

How white glue is used to glue two boards together edge-to-edge. Matching holes for dowels are first drilled into the board edges. Do this by clamping boards together, drilling for dowels.

To edge a table, a length of white pine is fitted, then holes are drilled through the edge and into the center section. Make certain the fit is flush. Glue is then applied with a brush.

Gluing up the legs. With the surfaces flat, (a jointer-planer does this job best) glue both halves together and clamp securely. Trim and sand all four sides after the glue has set.

Also brush glue on the end section and then locate it, by inserting a dowel through the stock and into the matching hole. Then drive all dowels home. For a perfect joint use clamps.

Silicone sealants are cream-like adhesives used as sealing compounds around bathtubs, sinks and washtubs. To use, a bead is pressed out of a tube. Formerly they would not accept paint, but now they are made so that the sealant can be painted to match the surrounding area.

Recorcinol is a powdered catalyst that is mixed with a liquid resin. It is an absolutely waterproof glue suitable for boatbuilding and all outdoor work. It can be used on leather, wood, concrete, plastics, laminates and chinaware. After thorough mixing, both surfaces to be glued are coated and clamped until the glue has dried. Temperature has a bearing on the setting time. At 70°, curing time is a minimum of ten hours, less at higher temperatures; it should not be used under 70°. Parts under stress should have the curing time doubled. Excess glue can be removed with cold water before it dries.

Powdered casein glue requires mixing and can be used for general woodworking. It is not waterproof, only water-resistant. It is a good glue for use on oily woods.

There is no single cement that can be used with *all* plastics. Every type of plastic on the market has its own particular characteristics and must be used only with a cement that is compatible. The dealer from whom you bought the plastic can advise you which cement to use.

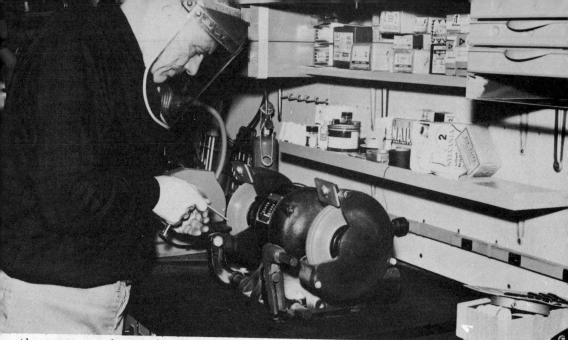

Always wear a mask or goggles when using grinder for any kind of work—stock removal or sharpening Grinder is used for sharpening drill bits, chisels, screwdrivers, plane irons and other similar edged tools.

Keeping Tools Sharp

It is the dull tool that is most apt to cause an accident. More effort is required to use it, it slips—and quick, where's the Band-Aid! If you watch a professional carpenter sawing wood, you will notice how effortlessly the saw seems to cut through the wood. That's because he is using a well-sharpened saw. Sharpening a crosscut or ripsaw is not at all difficult—just a bit tedious.

The first step in sharpening a crosscut saw is to lightly file the *tops* of the teeth to make very small flat areas on the points. These flat areas will serve as points of reference for the actual filing and sharpening. Use a small triangular file for sharpening. Clamp the saw in a vise between two strips of hardwood with the tip of the blade to the left. The bottom of the gullets (the "valleys") between teeth should be about 1/8-inch above the strips to avoid chattering during the filing operation.

Start at the blade tip and locate the first tooth that is set *toward* you, then rest the file in the gullet to its left. The file should be positioned so that it is in contact with the existing bevel of the tooth (about 45° to the blade as shown in the drawing). File down into the gullet until you have cut away half of the adjacent flat top made earlier. Skip the next gullet on the right and repeat the operation in the one after it. Continue until you reach the heel of the saw, skipping every other tooth.

Now remove the saw from the vise, turn it end for end and mount it in the vise with the handle at the left. This time however, you will start at the *right*—again the blade tip end—and place the file in the gullet to the right of the first tooth that is set *toward* you. This is the first gullet you skipped when you were filing the other side of the saw. Again position the file in the gullet to match the bevel of the tooth. File away until you have removed the remaining half of the flat area on top of the tooth. Keep up the good work until you reach the handle end at the left. Make sure you have not skipped any teeth. A trick used by old-time craftsmen is to blacken all the teeth with candle smoke. Sharpened areas show up most distinctly against the blackened areas.

A ripsaw is sharpened in much the same way except the file is used straight across the

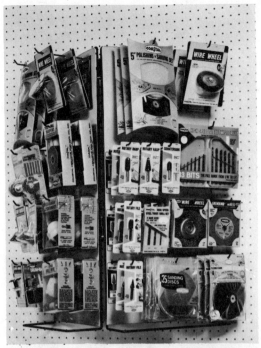

Sanding discs, wire wheel brushes, drills, burrs, and practically anything else you may need for sharpening, cleaning, wood or metal working can be found in attractive displays at hardware stores.

blade, at right angles to it as indicated in the drawing. Do not sharpen the entire saw from one side—sharpen every other tooth, the same as when sharpening a crosscut saw. Reversing the saw in the vise will tend to equalize the sharpening procedure and do away with any variations in the use of the file. Unequal sharpening will cause a saw to veer to one side during use. As a final step, check and tighten the bolts in the handle.

PLANE IRONS (BLADES) AND CHISELS

Two tools that must always be sharp if you are to get any work out of them are the plane iron and the chisel. To make it easy for you, both are sharpened the same way. The first step is to remove all nicks, straighten the edge and restore the bevel, best done on a power grinder. The grindstone should turn toward the plane iron or chisel; use the rest on the grinder to assure a flat even bevel, and keep the plane iron cool by frequent immersion in water to prevent softening of the steel. Move the blade from side to side to grind all parts of the bevel; this will

also serve to keep the wheel true. The resulting edge should be straight and at right angles to the sides of the plane iron or chisel. The correct grinding angle is between 25° and 30°. A good rule of thumb is to make the bevel just a little longer than twice the thickness of the plane iron or chisel. Too long a bevel is weak and can be nicked easily; a short thick bevel will not enter the wood easily.

After grinding, the next step is honing the plane iron or chisel on a whetstone (or oilstone). The honing or whetting angle is 30° to 35° (see the drawing). Apply oil to the whetstone so that the surface of the stone is kept moist during the honing operation. Place the bevel side of the iron against the stone with the back edge slightly raised; this will give you the extra 5° to 10° difference between the bevel angle and the whetting angle. Move the plane iron back and forth on the stone so that the angle between the plane iron and the stone always stays the same. Use enough oil so that the steel particles removed from the iron will not clog the stone.

After honing, you will note that the flat side of the iron has a barely visible wire or feather edge. This must be removed by making a few strokes on the stone with the flat edge of the iron or chisel *held absolutely flat against the stone*. The slightest bevel on this side must be avoided.

AUGER BITS AND DRILL BITS

Auger bits can be sharpened with a small file as shown in the drawing. Sharpen the spurs of the auger bit on the *inside* in order to preserve the diameter of the bit. Make sure both cutting edges are even; use a special auger bit file, shaped for this purpose; they are sold at most hardware stores.

Drill bits used for making holes in metal are made of tougher steel than auger bits and can only be sharpened on a power grinder. This is best done using a drill sharpening jig that holds the bit at the correct angle. Most twist drills of this type have a cutting angle of 59° as shown in the drawing. Both lips of a twist drill must be of the same length and ground to the same angle. The heel of the drill—the surface behind the cutting lips—must be ground away from the cutting lips at an angle of 12° to 15° as indicated in the drawing. Twist drills can be ground to the required angle and clearance on the side of a grinding wheel only if the grinding wheel is designed for this purpose; otherwise, the face of the grinding wheel must be used.

Unequal angles or lips will cause the drill to wobble, making an oversize hole.

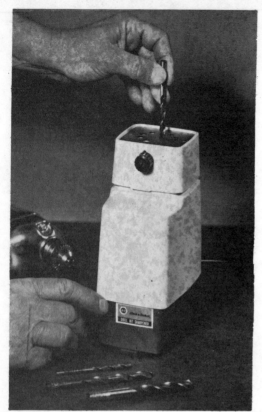

Drill sharpening device, made by Black & Decker, works like a pencil sharpener. Sharpens drills from 1/8th to 3/8th; just insert drill into proper hole and press to sharpen drills to factory specs.

Handy gadget for sharpening hedge clippers, garden shears and rotary lawn mower blades has tungsten carbide cutting surface; comes complete as a kit with an aluminum oxide sharpening stone.

A sharpening stone chucked into an electric drill can be used to sharpen spade drill bits. A little bit of metal taken off each side is enough to bring the bit back to its original sharpness; or use a file.

A special attachment is available for the grinder so that you can sharpen chisels and plane irons to factory newness. The photos shows a cutting tool for a lathe being ground to recommended angle.

TIPS ON GRINDING PLANE IRONS AND CHISELS

Plane marks will show less on finished work if the corners of the plane iron are rounded slightly.

Rocking the plane iron, or chisel, during the honing operation will produce a curved edge that will not cut well.

Finish the honing operation by taking a few strokes on a leather strip to produce a keener edge—barbers do it!

You can buy a jig for holding a plane iron or a chisel at the exact honing angle.

Know Your Saw Blades

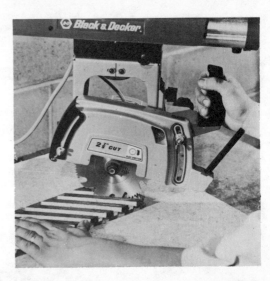

Having the right power saw blade on hand will expedite your work.

You will probably be doing most of your carpentry work—especially cutting—with power tools. It's the built-in accuracy of these tools, and their speed, that promotes good workmanship. However, you will not be taking full advantage of their capabilities unless you know something about the many types of blades available for your power saws. Five blades you should be familiar with, for use on the circular, bench, or radial arm saw, are described below.

The combination blade is used for both crosscutting and ripping. It is a general all-around blade suitable for cutting plywood, hardwoods and softwoods, and hardboard. It is the blade supplied as standard equipment with a new saw—portable or stationary.

The crosscut blade has finer teeth than the combination blade and makes a smooth cut across the grain of all woods, plywood, veneer and hardboard. Used extensively in building work to cut across 2x4s and other building timbers.

The rip blade has somewhat larger teeth than the combination blade and should be used when considerable ripping—cutting with the grain—is involved. However, this blade should not be used for ripping plywood as the crisscross pattern of plywood presents cross grain below the face veneer.

The hollow-ground blade makes the smoothest cut of all circular saw blades. It will cut with or across the grain with equal ease though as a rule it is mostly used for crosscutting. "Hollow-ground" refers to the fact that the teeth are not set outwards for clearance; instead, the body of the blade below the edges of the cutting teeth is thinner than the width of the teeth, thus gaining the necessary clearance to avoid binding.

Abrasive blades are made for cutting brick, masonry, metals and other hard materials. They consist of a fibrous body to which are bonded sharp particles of an abrasive. Each abrasive blade is specially formulated to cut a particular material. *Caution:* Wear safety goggles when operating power equipment—and especially when using abrasive blades.

There is also a variety of blades for use with the saber saw, or portable jig saw. The coarse-toothed blade is used for cutting thick wood; the fine-toothed blade for hardwood and plastic; the knife blade for cutting through linoleum, leather and rubber; the metal-cutting blade for use on iron, soft steel, brass and copper; the carbide-tipped blade for glass, ceramics and glazed tile; the taper, or hollow-ground blade for use when a very smooth edge on wood is desired. The same variety of blades is also available for the reciprocating saw as well as for the stationary jig saw.

Band saw blades also come in a variety of types. Their chief distinction is in the width of the blade; the smaller the width of the band saw blade, the tighter the circle it will cut. Band saw blades vary from 3/16-inch wide (will cut a circle one inch in diameter) to ¾-inch wide (will cut 3½-inch circles). Special blades for cutting plastics, aluminum and similar soft metals have a skip-tooth pattern. There are also blades for cutting hard metal. These are special high-grade steel blades accurately set for their purpose. The band saw must be set to operate at a lower speed than normal when it is used for cutting metal.

Eight Ways to Hang Shelves

Here are the key solutions to every storage problem you will ever have.

Right: cabinet with adjustable shelves to hold shoe boxes. The cabinet is on casters so that it can be moved out of the closet.

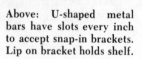

Above: U-shaped metal bars have slots every inch to accept snap-in brackets. Lip on bracket holds shelf.

Above: Shelf hardware made by Stanley, called, Flair, has matching hardboard shelves, brackets,

The greatest system ever invented for storage is shelving. Here are eight ways you can make supports that will keep your shelves where they belong.

1. The simplest is by means of a cleat at each end of the shelf. The cleats should be at least ¾-inch thick. Furring strips cut to the same length as the width of the shelf will do nicely. Nail or screw the cleats to the sides of the support.

2. The quickest way to build a shelf support is with brackets. These come in several sizes; the largest will support a 12-inch wide shelf. When installing brackets, always mount the longer leg against the wall. The short leg supports the shelf. The chief drawback with brackets is that the shelf has a tendency to be "springy," especially if the shelf is loaded with heavy items. Shelf support brackets with gussets are more sturdy, or you can reinforce these brackets by adding your own gussets. Cut a length of strap iron and bolt or weld it to the legs of the bracket.

3. The professional way to mount a shelf is by cutting dadoes into the opposite sides of the framework. The dado should be just a bit wider than the thickness of the shelf so the shelf will not bind as it is slid into place.

4. Angle irons can be used to support a shelf. Use two at each end, one in front and one in the rear for a total of four for each shelf. Use an extra angle iron in the middle if the shelf is very long.

5. Instead of angle irons, you can use 3/8-inch dowels. Drill holes for the dowels, two at each end. Bevel the dowel ends slightly before driving them in. To prevent the shelf from "rocking," make sure that the dowels are all at the same level.

6. Shelves which must carry great weight can be adequately supported on long, threaded ¼-inch steel rods. This is a good method for shelves in the basement where the ceiling joists are accessible and the rods can be placed as close together as needed for the heavy loads. The rear edges of the shelves are supported by 1x2-inch furring strips as shown in the drawing. Spacing of the shelves can be adjusted by loosening and running the nuts up and down as required.

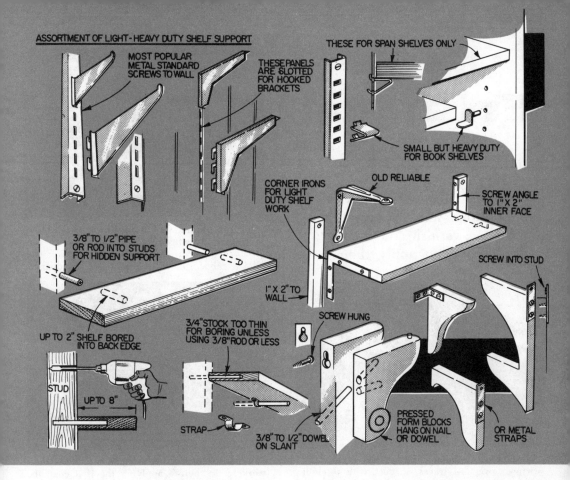

7. Short shelves can be supported by securing them with screws driven from the outside of the supporting framework. Use fairly long screws as you will be driving them into the end grain of the wood.

8. A shelf support made of 1x3 furring strips is constructed by cutting one end of the support at a 45° angle and notching the upper part of the support to fit the opposite end of the 1x3. This is a good system for supporting heavy loads in such places as the garage.

Of course there are store-bought ways of putting up shelves. One of the simplest systems consists of two metal tracks which are screwed to each side of a cabinet or support. The tracks have a series of slots at one-inch intervals. Clips, part of the package, are inserted into the slots to support the shelving. These tracks are made in several styles. One type is designed to fit flush within a groove cut into the supporting structure. Another type is surface mounted and projects slightly, about a half-inch, from the cabinet or supporting sides.

Some of these tracks, instead of using clips to support the shelves, use a system of brackets. The brackets are made in sizes to fit shelves of various widths up to 12 inches. One type is locked into place with a locknut; another type is secured with a friction fit, requiring a slight hammer blow to lock it in place. Some brackets can be adjusted to slant 15°, 30° and 45° from the horizontal for storing magazines, books and records. Still another system uses a series of keyhole openings instead of slots as part of the support system. And if styling is a factor, you can even get brackets that will support the shelves from the top, instead of from the bottom.

The shelves themselves can be purchased prefinished in solid wood, veneered, hardboard covered with plastic laminate, and in finishes to match practically any type of wood available. Widths vary from six to twelve inches, length is usually 48 inches. Two brackets spaced six inches from each end are sufficient to support these shelves for normal loads.

It's the Finish That Counts

What you should know about filling, staining, Japan colors, shellac, varnish and lacquer; and how to apply them.

The final sanding step in finishing a piece of fine furniture is with extra-fine sandpaper. Wrap the sandpaper around a block of wood or use 3M's rubber block to hold the sandpaper.

Next comes staining, after you have carefully wiped away all traces of dust with a cloth lightly dampened with paint thinner. Apply the stain with and across grain to force it into wood.

After the stain has been wiped off, allow the work to dry before applying the first coat of shellac, varnish, or lacquer. Allow ample drying time between all coats. Finish by waxing.

No matter how much work you have put into building a piece of furniture, or how painstakingly you have made the joints, unless the finish is well-nigh perfect your work may be for naught—or that is what your relatives will mutter under their breath when you ask them to dinner so they can admire your workmanship.

Before starting any finishing operation, check all joints. Make sure they are tight, securely glued, doweled or screwed. Clean away any excess glue from around the joints. Next comes sanding—a job to be done thoroughly, carefully and conscientiously. A power finishing sander can be used, but do not depend on it to do all the work. There are crevices and curves which can be done only by hand. Always sand with the grain. The orbital sander has the advantage of greater cutting speed than the straightline sander, but it does half its sanding across the grain, for this reason you should finish with a straightline sander or by hand.

Always use a fine grade of sandpaper in these operations. Use the coarser grades only for initial sanding and to get rid of deep scratches and gouges.

For hand sanding flat surfaces, hold the paper around a block of wood, or use a commercial sandpaper-holding device, available at hardware stores. Wide curves can be sanded by holding the paper in the palm of your hand; narrow interior curves are best sanded by wrapping some paper around a dowel or a suitably curved piece of wood which more or less matches the curve of the wood you want to sand. Exterior curves, such as the rungs of a chair, can be sanded by pulling a narrow strip of sandpaper back and forth around the curved piece, somewhat like polishing your shoes with a rag. Narrow crevices can be sanded by folding the sandpaper into two or more thicknesses. The sandpaper will wear most rapidly right at the crease, so keep refolding it to expose fresh sandpaper.

FILLING

After sanding, the next step is either staining or filling, depending upon the wood. Filling is the process of filling up the pores of the wood. Some woods have large open pores, such as oak, ash, chestnut, elm, mahogany and walnut. The filler makes the indented areas of the wood—the pores—flush with surrounding areas. Most fillers are of the paste type, used for open-pore woods; liquid filler is used for close-grain woods such as maple, cherry, beech, birch, and gum.

Paste fillers are available in many shades and oil colors can be added to match any desired hue. The filler, which comes in a can, should be thinned slightly with turpentine, benzine or a paint thinner, to the consistency of heavy cream—just thin enough to brush. Too much thinning will defeat the purpose of the filler, which is to fill the grain of the wood and not just to impart a texture to the wood. Use a fairly wide brush to spread the filler thoroughly over the surface of the wood. You can brush with the grain or across it; the important thing is to make certain that *all* areas are covered. Let dry for 10 to 15 minutes or until the filler has begun to lose its luster and become dull. Now make up a pad of burlap or a comparable coarse material and rub across the grain to remove excess filler and at the same time rub the filler into the grain or pores of the wood. Make sure you go over all areas of the wood.

When you are satisfied that all areas have been covered, make a wad of a softer material, such as cheesecloth, and again wipe the surface, but this time wipe *with* the grain. Keep up the good work until the rag shows no more pickup of filler. Rub the wood with your fingertips. If no color appears on your fingers, then you have done a good wiping job. The wood should now exhibit a soft sheen when viewed by reflected light.

Make sure that all irregular surfaces, crevices and corners have received the filler. But suppose there are some areas on which you find it impossible to wipe away the surplus in time and it dries hard. Just dampen a rag with some paint thinner and go over these areas and finish off with the soft rag. Every bit of filler not in the pores of the wood must be removed before proceeding to the next step. You can get rid of excess filler in corners and crevices by wrapping a rag around a dowel sharpened to a point. The covered point will fit into these normally hard-to-get places, but keep changing the position of the rag.

LIQUID FILLERS

Liquid fillers for close-grain woods usually have a varnish or lacquer base and contain relatively little solid matter. Brush on the filler, the same as you would varnish or paint; avoid a heavy coat. After the filler is dry, sand to get all excess off the surface of the wood. Shellac is sometimes used as a filler. The technique is to use two thinned-out coats of shellac, with a light sanding between coats. Shellac is thinned with alcohol, *not* a paint thinner.

STAINING

Always wait at least 24 hours after applying a filler before staining wood. It is possible to apply stain *before* the filler operation. However staining after filling is preferable, especially if the filler does not exactly match the wood; stain tends to cover the filler, thus making the match.

Staining has two functions: it emphasizes the beauty and grain of the wood, and it brings the wood to the desired color. For a so-called natural finish, use a light stain that will not affect the color of the wood, or you can bypass the staining operation completely for a true natural finish.

Penetrating oil stains consist of dyes dissolved in a carrier—benzine or turpentine. They come ready to use, provided you can find one in the color you want. This should be no problem as they are made in dozens of colors for all popular woods.

Apply the stain to the work with a rag or a brush, whichever you prefer. Apply generously, with the grain and across it. Let it soak in for about 15 minutes and wipe away excess with a clean rag. Don't allow the stain to dry as it will cake and affect the final finish. The final wiping should always be done with the grain of the wood.

You can also buy **water-soluble stains**. These tend to accentuate grain lines and are sometimes desirable for this purpose. They come in powdered form and are applied by brush or cloth. You can alter the color somewhat by adding more water to make the stain lighter, more powder to make the stain darker. Excess stain can be removed from the work by wiping with a rag dipped in water. It is best to apply the stain in two coats—the first should be a light one, the second one to match the desired color.

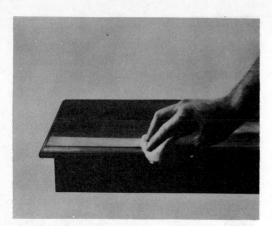

For a light staining, wipe off the stain after only a few minutes of drying time. Never allow the stain to dry completely. Dried stain should be removed with a rag and some paint thinner.

Because you do your own mixing of these stains you can get practically any color in the rainbow. But water stains tend to raise the grain on wood. After applying a water stain and letting it dry overnight, go over the entire surface of the work with fine sandpaper, rubbing with the grain, to remove all the tiny slivers of wood that have appeared.

These stains, because they are "clear" and do not have suspended particles of color the way oil stains have, produce a clean, sharp appearance, much more impervious to fading by light than oil stains. *Caution*: Water stains cannot be used on wood that has been previously finished, even if all of the old finish has been removed.

To eliminate the need for sanding caused by raising of the grain, there is another type of stain on the market called N.G.R.—non-grain-raising stain. These have most of the advantages of water stains, but they are not as colorfast. Because they are sold ready-mixed, color selection is limited, but they can be intermixed to produce other hues. These stains, mixed in alcohol, dry faster than the water type and you can proceed with a finishing operation a half-hour, or even less, after applying the stain. Another important advantage is their ability to penetrate a lacquered or shellacked surface and stain the wood beneath. Like the water stains, alcohol stains are "transparent" and leave a clean, sharply defined grain pattern after application.

Japan colors are pigment-type stains and are used for painting, glazing, blending and toning as well as for staining. They are ground 6 to 12 times—the finer the grind, the better the results, and the more expensive. Used on raw wood, they tend to hide the grain rather than accentuate it. When used in a strong concentration, japan colors act just like a paint. They are available in all wood shades as well as in black, white, and green; when mixed with a lacquer, they give it a color to form a sort of toner; when mixed with sealer they form a colored glaze. Japan colors should not be confused with japan dryer, a fast-drying varnish mixed with a stain to hasten the drying process. Don't use an excess of dryer to unduly hasten drying time. Too much japan dryer will gum up a stain and cause alligatoring.

SHELLAC

Shellac is among the oldest of all finishing materials. It is made from an insect called lac, which inhabits parts of India and Ceylon. This insect lives, breeds and dies in trees. It exudes a hard substance which eventually covers the insect and causes its death. Dead insects on the twigs of the tree are harvested, crushed and ground into granules, further refined into flakes and exported in 100-pound sacks to shellac manufacturers. For many years, most furniture was finished with shellac. Nowadays, assembly-line furniture construction calls for the use of fast-drying lacquer, but shellac still deserves its popularity—with aging, it gains a beauty and mellowness that cannot be matched by other finishes.

Shellac is sold by the "cut," which indicates the amount of shellac by weight dissolved in a gallon of alcohol. The standar cut of shellac is the five-pound cut (five pounds of lac dissolved in one gallon of alcohol). A three-pound cut contains three pounds of lac dissolved in a gallon of alcohol. The standard cut of shellac is the five-pound cut (five pounds of shellac dissolved in one gallon of alcohol). A three-pound cut contains three pounds of shellac dissolved in a gallon of alcohol. The standard cut of shellac is the five-pound cut (five pounds of shellac dissolved in one gallon of alcohol. A three-pound cut contains three pounds of shellac dissolved in one gallon of alcohol. Obviously, a three-pound cut should cost less than a five-pound cut—and it does. It is best to buy shellac in a five-pound cut and thin it down yourself as required. The thinner the shellac, the easier it is to brush on. As in painting, two thin coats are better than one heavy coat. The table shows the amount of alcohol needed to thin shellac for various cuts.

QUANTITIES FOR REDUCING SHELLAC	
Shellac	Alcohol Added
5-lb cut to 3 lb	3½ pints to 1 gal shellac
5-lb cut to 2 lb	1 gal to 1 gal shellac
5-lb cut to 1 lb	2 2/3 gals to 1 gal shellac
4-lb cut to 3 lb	1 qt to 1 gal shellac
4-lb cut to 2 lb	¾ gal to 1 gal shellac
4-lb cut to 1 lb	2 1/8 gals to 1 gal shellac

There are two types of shellac sold: *orange* and *white*. Orange shellac should be used on dark woods such as walnut and mahogany; it will not affect the color of the stained surface. White shellac (it really isn't white, but more of a tan color) has been lightened by adding a bleaching agent to orange shellac. It can be used on dark surfaces but is generally applied on light-colored woods, or bleached woods, where it is desirable to leave the color of the wood as is. White shellac is much more perishable than orange shellac because of the bleaching agent. It has a limited shelf life, about ten months, and should be discarded if it is more than a year old. Some manufacturers date their containers, assuring the purchaser of fresh shellac.

Shellac serves as an excellent sealer on fresh wood or stained wood. It also serves as a sealer before painting plastered walls. Paint and enamel, used over a shellac-sealed surface will not soak in, thus making the paint, or enamel, go farther. When using shellac as a sealer, thin it down to a two-pound cut. Avoid a heavy coat of shellac as it may cause the final finish to crack or alligator. Shellac dries very quickly, so you can apply a second coat after a two-hour wait. Three or four coats of shellac, thinned down to a two-pound cut, make a beautiful, durable finish. It is not necessary to sand between coats, as each succeeding coat of shellac, because of its alcohol content, tends to flow into and combine with the layer beneath. After the last coat of shellac has been applied, wait at least 24 hours and rub down the surface of the work with fine steel wool and paste wax. Polish with a terry-cloth towel for a beautiful, semi-lustrous finish.

VARNISH

A varnish consists of a resin or gum, which is the "body"; tung oil or linseed oil, which is the "vehicle"; a thinner, usually turpentine; and a dryer to hasten the drying action after the varnish is applied. Colors range from clear to a dark brown. "Spar" varnish includes an additive that makes it impervious to salt water and sun. This varnish, obviously, is much favored by boatmen, but you can use it just as well for outdoor furniture. A good varnish will expand and contract with the furniture on which it is applied without cracking.

Varnish should be applied in a working area which is as dust-free as possible. Use a fairly soft brush, as wide as you can comfortably handle. If you are going to apply the varnish to bare wood, thin the first coat three to one—one part of varnish to three parts of turpentine. Use this as a sort of sealer. Varnish dries much more slowly than lacquer, shellac or even paint. You should wait at least a day, or even more, before applying the next coat. Sand lightly with fine sandpaper between coats and wipe off the dust with a cloth dampened with turpentine. Run your fingertips over the surface of the work—they will tell you better than your eyes if you have missed any areas when sanding.

Second and third coats should be applied just as the varnish comes from the can. A trick employed by some finishers is to partially immerse the can in a saucepan of warm water. This tends to make the varnish flow more readily. Boatmen do this during cold weather.

Polyurethane varnishes produce a very tough surface for exterior and interior work; they are highly resistant to alcohol, grease, most acids and to hot and cold water. Their slightly higher cost is well worth it.

A brush once used for paint should never be used for varnish, no matter how well it has been cleaned. Best bet is to keep on hand one or two brushes to be used exclusively for varnishing.

After the final coat of varnish has been applied and allowed to dry thoroughly, you may be somewhat disappointed with the result. The varnish may look too "shiny," or too many dust specks are apparent on the surface. This is when you bring out the beauty of a varnished surface with pumice and a bit of elbow grease. Make a felt pad about three by four inches and about ¼-inch thick; an old felt hat is ideal for this purpose. Place a fine pumice stone in a shallow tray and add a few ounces of light machine oil. Soak the pad in the oil, rub it over the pumice stone to pick up some of the fine abrasive particles and rub the felt pad, charged with the pumice, over the work. Use moderate pressure, let the pumice and oil do the work. Keep rotating the pad in your hand to present a new cutting surface and recharge it by rubbing it over the pumice stone when the cutting action seems to slow down or disappear. Keep up the good work, making sure you do not miss corners and crevices. Wipe off the muck with a soft cloth and you will be pleasantly surprised at the soft sheen your work now exhibits. Look for missed spots and go over them.

You can use water instead of oil, but water and pumice will cut faster than oil and pumice, so use a little less pressure. Don't allow the water to remain on the surface too long as it may penetrate and affect the varnish. Wipe with a water-dampened cloth to remove the residue of water and pumice. Whether you use water or oil, a hard paste wax provides the final finishing. Make a pad of terry cloth, dampen it slightly with water, rub it into the can of wax—floor wax or automobile wax is fine—and apply diligently to your work.

LACQUER

If you have a spraying outfit, there is no easier way to finish a piece of furniture than with lacquer. Even without a spray, you can still use lacquer as there are plenty of lacquers especially made for brush application. The same caution about a brush and varnish applies—never use a brush that has been used for paint *or varnish* for a lacquering job.

Lacquer is readily identifiable by its banana-like odor. It dries fast and does not conceal the grain of the wood the way other so-called clear finishes do. Never apply lacquer over a painted surface; it will lift the paint, somewhat like a paint remover. This is because of the compounds that go into its manufacture. All lacquers contain five main ingredients; a synthetic or a natural gum, nitrocellulose, solvents, softeners and some sort of thinning agent. All these ingredients may be combinations of other chemicals as well.

When spraying lacquer, give the work at least three coats. A heavy spray deposit is bound to give you trouble in the form of runs. A run in lacquer cannot be picked up the way you can pick up a paint run—with a brush. The only thing you can do is cuss a bit, wait until the lacquer dries thoroughly and sand the run down. So it's best to use a thin spray. Lacquer designed for spraying is much thinner in consistency than lacquer designed for brushing, but you can thin brush-type lacquer for spraying by adding lacquer thinner. Do not use any other solvent for this purpose.

There are many special-purpose lacquers. A few of these are described.

Flat lacquer dries to a dull, mat finish. It can be mixed with a high gloss lacquer to produce any desired in-between sheen.

Clear, glossy lacquer is transparent, slightly amber in color and dries to a high gloss. After a few minutes of drying time, it is practically dustproof and can be sanded after two hours. Note: sanding between coats is not necessary when applying lacquer as each succeeding coat will melt into the preceding one. However, sanding is sometimes necessary to remove dust particles that may have settled on the surface during the drying process.

Water-white furniture lacquer is similar to the above, except it is clear as water and is generally used on bleached or natural-finish woods so that no darkening of any kind is effected.

Shading lacquer comes in many colors such as black, brown, blond, white, wheat, etc. It is usually applied by spray only after several coats of clear, glossy lacquer have already been applied.

White, undercoat lacquer is a white, opaque lacquer frequently used as an undercoat for colored lacquers. It can be used as a sealer and can be tinted to the shade of the finish coat of lacquer. It should not be used when a natural wood finish is desired.

Lacquer enamels are almost invariably applied by spraying and come in literally scores of colors. They dry to a hard, durable surface that can be cleaned and washed without injury to the lacquer coat. These lacquers can be used as a paint substitute.

Lacquer thinners. Only the thinner recommended by the particular maker of the lacquer you are using should be employed as a thinner. Thinners vary in chemical make-up and the wrong thinner may adversely effect the lacquer you are using. All lacquer thinners are water-white in appearance. When used as an undercoat or as a sealer, lacquer should always be thinned before application.

Retarders. Sometimes your lacquer finishing job may "blush." This blush, a cloudy-white appearance, is due to absorption of water from the air as the lacquer is drying. In order to prevent this, a special retarder-thinner is used which permits the lacquer to dry more slowly and thus minimize the absorption of moisture. Best bet is not to apply lacquer on warm, humid days.

Paints and Enamels

How to select the right type for indoors or outdoors. How to prepare for painting; professional methods of applying paint.

Painting is 90 per cent preparation and cleanup. If you are planning to do any painting of any kind, the very first step before applying the brush is to lightly sand and clean all surfaces.

Get a brush to fit the job and quite often you will find that two brushes of different widths will expedite the job—even though you will have two brushes to clean. Keep a clean rag for drips.

More finishing is done with paint than with all other materials combined. In recent years there has been a virtual revolution in the paint industry. Paints that can be thinned with water, paints that dry with a speckled finish, paints that are mixed with a hardener to make an epoxy finish, and paints that dry with a metallic finish are just a few of the many different products now on the market.

Latex or water-soluble paints appear to be most popular with the homeowner. They have many qualities to recommend them: they dry in an hour, ready for a second coat; they have good covering qualities; brushes and rollers can

EXTERIOR PAINTING GUIDE ITEMS LISTED BELOW ARE BEST PROTECTED USING PAINTS LISTED HERE → ◎ MEANS PRIMER OR SEALER COAT MAY BE NEEDED	ALUMINUM PAINT-EXT.	ASPHALT EMULSION	AWNING PAINT	CEMENT BASE PAINT	EPOXY PAINTS	EXTERIOR MASONRY LATEX	EXTERIOR CLEAR FINISH	HOUSE PAINT-LATEX	HOUSE PAINT-OIL BASE	LATEX TYPES	METAL PRIMER	PORCH AND DECK ENAMEL	PRIMER OR UNDERCOAT	ROOF COAT-SURFACING	ROOFING CEMENT	SPAR VARNISH	TRIM PAINT	TRANSPARENT SEALER	WATER REPELLENT PRESERVATIVE	WOOD STAIN
ASBESTOS CEMENT						○		○	◎			○						○		
BRICK WORK — — — — —	◎			○		○		○	◎			○						○		
CANVAS – AWNINGS			○																	
CONCRETE BLOCK	○			○		◎		○	○			○						○		
CONCRETE FLOORS					○					○	○									
COPPER SURFACE																○				
GALVANIZED METAL	◎							◎	◎	○						○	◎			
IRON WORK	◎							◎	◎	○							◎			
METAL WINDOWS	◎							◎	◎	○							◎			
METAL ROOFING								◎	◎											
METAL SIDING	◎							○	◎	○							◎			
NATURAL WOOD TRIM							○									○				○
STUCCO WORK	○			○		○		○	◎	○		○						○		
SHUTTERS AND TRIM								◎	◎	◎							◎			
TAR FELT ROOF		○												○	○					
WOOD SIDING	○							○	◎	○		○								
WOODEN WINDOW FRAMES	○							◎	○	○		○					◎			
WOOD SHINGLES																			○	○
WOOD FLOORING												○								

Painting your fence? First, remove all the loose and scaling paint. Then tighten all joints and make certain rails fit snugly. Drive protecting nails below the surface and seal with shellac.

This spray outfit, put out by Black & Decker is a great time-saver when it comes to painting items such as shutters and screens, even though spraying may require more than one coat.

be cleaned with water and soap. A couple of disadvantages: at present you cannot get a high-gloss latex enamel in all colors and the actual thickness of a latex paint coat is somewhat thinner than a single coat of oil paint.

INTERIOR PAINTING

Latex paints are also available for exterior use on clapboard and trim. Because of their water-solubility, they can be applied during damp weather and even over a surface that is slightly damp.

Undercoats. When painting raw wood, an undercoat is usually recommended. The undercoat serves as a sealer and forms a hard, tough base for the paint. The disadvantage of using the same paint for the undercoat as for the finish coat is that quite often spots are missed because there is no color difference. You can use a flat paint for the undercoat or even white lead thinned with turpentine. *Caution*: Do not use white lead or any lead-containing paints in a child's room or where a child may have access. Lead is a poison when taken internally. Shellac can also be used as an undercoat and a sealer. Open-grain woods such as oak, walnut and ash have such large pores that even three or four coats of paint will not hide the grain. These woods should always be sealed with an appropriate undercoat.

EXTERIOR PAINTING

For interior work, there are three types of paints generally used: a flat finish for ceilings and walls; a semi-gloss or a full gloss for the woodwork in a room; and a deck-type enamel if the floor is to be painted. The last mentioned is a tough, varnish-based paint which can take floor traffic without undue wear. Flat and semi-gloss paints come in latex or alkyd, but floor enamel is an alkyd type. (There are latex paints made for floor use but they will not wear as well as the alkyds.) The enamels or glossy paints will take more abuse, scrubbing and washing than flat paints.

If in doubt about which type of paint to use, consult the chart. Note that a star indicates that a primer or sealer should be used before applying the finish coat, unless the surface has been previously painted and is in good shape.

After you have selected the proper paint, preparation is next in order. Any successful paint job consists of 90 per cent preparation and only ten per cent wielding the paint brush. Preparation consists, first, of getting all your supplies together (brushes, paints, thinners, dropcloths, ladders, rags, sandpaper, scraping tools, steel wool, rollers, pans, planks, hand tools, etc.). If you are going to paint a room, remove all furniture, or place it in the middle of the room adequately covered with dropcloths; remove all light fixtures, or drop them from the

Some horrible examples of paint failure. No. 1 was caused by painting over a surface that was not dry; No. 2 paint flaking under a soffit, insufficient ventilation in attic above is the cause; No. 3, a paint of poor quality, it just wouldn't stick; No. 4, painting over a poorly prepared surface; No. 5, alligatoring caused by applying a second coat before first dried; No. 6, bleeding knot. The answer is preparation before painting!

INTERIOR PAINTING GUIDE

INDOOR SURFACES LISTED BELOW ARE BEST FINISHED WITH PAINTS LISTED HERE

◎ MEANS PRIMER OR SEALER MIGHT BE NEEDED FIRST

Surface	ALUMINUM PAINT	CEMENT BASE PAINT	CLEAR POLYURETHANE	EPOXY PAINTS	FLOOR VARNISH	FLOOR PAINT AND ENAMEL	FLAT PAINT – ALKYD	FLAT PAINT – LATEX	GLOSS ENAMEL – ALKYD	INTERIOR – VARNISH	METAL PRIMER	RUBBER BASE PAINT-NO LATEX	STAIN	SHELLAC	SEMI-GLOSS – ALKYD	SEMI-GLOSS – LATEX	SEALER OR UNDERCOAT	WOOD SEALER	WAX – LIQUID OR PASTE	WAX – EMULSION
ASPHALT TILE FLOORS																				○
CONCRETE FLOORS				○	○	○						○	○						◎	◎
DRY WALLS							◎	◎	◎					○	◎	◎	○	○		
HEATING DUCTS	○						◎		◎		○	○			◎	◎				
KITCHEN AND BATH WALLS									◎			○			◎	◎	○			
LINOLEUM				○	○								○						○	○
METAL WINDOW FRAMES	○						◎	◎	◎	○	○				◎	◎				
NEW MASONRY WORK		○					◎	◎	◎			○			◎	◎				
OLD MASONRY WORK	○	○	○				○	○	○			○			○	○	○			
PLASTER WALLS – CEILINGS							◎	◎	◎			○			◎	◎	○			
RADIATORS – HEAT PIPES	○						◎		◎		○	○			◎	◎				
STAIRCASES – TREADS – RISERS			○		○	○						○	○	○				○		
STEEL SURFACES – CABINETS							◎	◎	◎	○	○				◎					
WALL BOARD							◎					○			◎	◎	○			
WINDOW SILLS – FRAMES							◎	○	◎								◎			
WOOD CASING – TRIM			○			○	◎		◎	○		○	○	○	◎		○		○	
WOOD PANELING			○			○	◎		◎	○		○	○	○	◎			○	○	
WOODEN FLOORING			○	○	○	◎							○	◎					○	◎
VINYL – RUBBER – TILES																			○	○

Left: A three-inch wide brush is about right for painting interior trim such as door jambs and baseboards. Use a wider brush for the walls. Remove outlet and switch plates before painting.

Always cover the lid of a paint can with a cloth before hammering it closed. You will avoid getting an eyeful or messing up clothes.

ceiling; remove all outlet and switch plates. Scrape away all loose paint, sand to a feather edge and apply a prime coat to freshly patched areas (use spackle for patching). Remove all hardware from doors, and windows; vacuum thoroughly to remove dust and debris; wipe all areas to be painted with a cloth dampened with water if you are going to use a latex paint and a cloth dampened with paint thinner if your paint will be an alkyd paint. Now you can start to paint!

The place to start is the ceiling. The best way to paint a ceiling in order to avoid needless ladder shifting is to use two ladders and span them with a sturdy plank. Mentally divide the ceiling into sections, say thirds the long way and thirds again the short way. Do one section at a time, using the plank as a walkway. You'll

Brighten up your basement with white paint. Over a rough surface such as cinder blocks, you may have to apply two coats. Use a four-inch brush. Cover area with newspapers to catch drips.

Concrete basement floors can be painted. The first step is vacuum the area to be painted. Do four-by-four-foot section at a time. Loose and scaling paint is removed by vigorous scraping.

When you paint the basement floor, paint a "base board" about four inches high on the wall. And do this before you have painted yourself into a corner! Gray floor enamel is the best.

Steps should present no particular problem even if they are in use. You can either paint one-half at a time or else paint alternate steps and caution the family about your clever technique!

be pleasantly surprised at how much faster the work goes using the plank-and-ladder system instead of constantly shifting a single ladder.

When painting the ceiling, use a roller or as wide a brush as you can comfortably handle. Pros generally use a six-inch brush, but you may find this a bit too wide; a four-inch brush will do the job nicely. Hairline cracks in the ceiling will most often be covered by the new paint. Wider cracks should first be filled with spackle or patching plaster. Cracks more than 1/8-inch wide should be cleaned out with a beer can opener and then filled with spackle or plaster. Sand after drying and then paint.

The next step is to paint the woodwork and trim. If you are really fastidious, you will remove not only locks and hardware from the doors, but also the hinges, and paint the doors separately. Old paint can be removed from hinges by soaking them in a lye solution—a ta-

blespoonful of lye to a quart of water. *Caution:* Be careful when working with lye, avoid spatters, and mix only in a glass, plastic or steel container—*never use an aluminum container for mixing lye*. Let the hinges soak in the solution overnight, remove them with a tong, rinse in cold water, let dry, and spray with clear lacquer or coat with varnish to keep the bright finish of the hinges intact.

All trim (woodwork, doors, baseboards, molding, etc.) should be lightly sanded and wiped clean before painting. If your floors are carpeted wall-to-wall, use a piece of stiff cardboard to protect the carpeting from the brush as you paint the baseboards. Use dropcloths to protect all other areas of the room. It is best to paint the trim of the room first, and then the walls. It is easier—or maybe it just seems so—to paint a straight line on the walls adjacent to the trim rather than the other way around.

INDEX

Adhesives 106
    contact cement 106
    epoxies 106
    latex-base adhesives 106
    mastic adhesives 106
    outdoor furniture 106
    plastic cement 106
    powdered casin glue 107
    Recorcinal 107
    rubber base cement 106
    silicone sealants 107
    white glue 106
Alligatoring 116
Beading 38
Bits 27, 38
    drills 27
    router 38
Bolts 97
    eye bolts 97
    fillister head 95
    flat head 95
    hangar 95
    J-bolts 97
    lag bolts 97
    machine 95
    Molly 95
    oval head 95
    round head 95
    sizes 97
    toggle 97
Box joint 72
Braces 17
    carpenter's 17
Bridging 78
Casters 65, 67
Ceilings 100
    acoustical tiles 102
    full suspension 102
    Integrid 100, 101
    light fixture 103
    remodeling 103
    tile 100
Chalk 22
Chamfering 38
Chimney 83
Chisels 18
    cold 18
    double header 25
    rip 25
    wood 18
Circles and arcs 22

Clamps 20
    C-clamp 20
    jiffy 20
    pipe 20
    wood, adjustable 20
Closet, cedar 104
    basement 104
    room divider 104
Compass 22
Concave cuts 38
Countersink 24
Curves 44
Dado 30, 38
    plain 70
    stopped 70
Dadoing 41
Doors 85, 87, 123
    lock 86
Dovetail 38, 70
    lap 70
    miter 70
Drawer construction 72
Drill press 45
    attachments 45
    drum sander 45
    files 45
    rasps 45
Drills 17
    attachments 26
    electric 17
    hand 17
    quarter-inch 26
    variable speed 28
    Yankee push 17
Dry wall construction 60
Equal measurements, making 22
Enamel 119
Epoxies 106
Files 24
    flat 25
    needle 25
    round 25
    semi-round 25
    smooth end 25
    triangle 25
Filling 115
    liquid 115
    paste 115
Floor joists 78

Floor, pegged 96
Formica 61
Framing 81
    ceiling 81
    walls 81
Framing 76
    balloon 76
    ceiling 81
    house 76
    platform 76
    wall 81
Furniture, gluing 20
Girders 78
Glazing 116
Grinder 40
Grooves 13, 38, 41
Hammers 6
    claw 6
    ripping 6
    short-handled sledge 6
Hardboard, perforated 60
Hinges 123
Hinges, mortising 38
Jig saw 47
    saber blades 47
    table 47
Jointer 40, 49
    table 49
Joints 70, 73, 76, 78
    box 72
    bridle 73
    copped 74
    cross-lap 73
    cross-over 73
    edge to edge 73
    fish plates 73
    joining plates 73
    lengthing 73
    lapped 73
    miter 74
    scarf 73
    splayed lap 73
    V splice 73
    splined 73
    three-way 73
Joint tenon 70
    dovetail 70
    stub tenon 70
    T joint 70
Joint knife 22

Joists 78, 81
   ceiling 81
   floor 78
   size 81

Kitchen countertops 92, 93
   edge molding 93
   plywood 92
   vinyl 92

Kitchen cabinets 88, 89, 91
   base 89
   flush 89
   recessed 89
   sliding doors 89
   steel 91
   wall 89

Knife 22
   joint 22
   putty 22
   utility 22

Lacquer 118
   clear 118
   enamel 118
   flat 118
   furniture 118
   retarder 118
   thinner 118
   undercut 118

Lathe 40, 50, 81
   face-plate turning 50
   spindle turning 50

Legs 73
   attaching 73

Levels 16
   carpenter 16
   masonry 16
   monovial 16
   pocket 16
   spirit 16
   torpedo 16
   vial 16

Lintels 81

Lock 86
   door 86
   mortise 86
   rim 87
   spring latch 87
   marking gauge 24
   mason's line 22

Measuring devices 15
   folding wood ruler 15
   folding zig-zag ruler 15
   steel tape 15

Micarta 61

Miter box 18, 75

Miter gauge 41
Mitering 42
Miter joint 72

Molding 74
   ceiling 74
   floor 74

Mortise 70

Nail set 22

Nails 94
   common 94
   penny size 94
   special purpose 94

Nippers 23

Ogee curves 38

Outlets 56

Painting 116

Paints 121, 123
   alkyd 121
   deck type 121
   flat finish 121
   full gloss 121
   latex 121
   semi-gloss 121
   trim 123
   undercoat 121

Paneling 56, 58
   butt joint 58
   contemporary 58
   estimating 56
   hardwood 58
   installation 56
   soft wood 58
   solid wood 58

Perforated hardboard 61

Plane irons, sharpening 109, 110

Planes 13
   bench 13
   block 13, 37
   electric 37
   jack 13, 37
   jointer 13
   trimming 13

Plastic laminates 60, 63
   sheets 61

Pliers 22
   cable cutter 23
   common slip joint 23
   heavy duty 23
   locking (vise grip) 24
   needle nose 23
   parallel jaw 23
   round nose 23

   square nose 23
   wire cutter 24

Plumb bob 22

Plywood 54
   DFPA 54
   exterior 55
   furniture 54
   interior 55
   lumber core 54
   paneling 56
   quality 54
   strength 55

Posts 78

Power tools 26, 67

Pry bar 22

Putty knife 22

Rabbets 13, 38

Radial arm saw 42

Rafters 82
   common 82
   cripple jack 82
   hip jack 82
   valley jack 82

Rasp 24

Ripping 41

Roof frame 82

Roofing square 82

Router 38

Router guide 38

Sandpaper 24

Sander 32, 46
   belt 32, 33, 46
   disc 46
   finishing 32
   orbital 32
   power 33
   straightline 32

Sanding 98, 114
   abrasives 98
   emery 98
   flint 98
   garnet paper 98
   grades 98, 99
   power sander 99
   sandpaper 98
   silicon carbide 98

Saw blades 111
   abrasive 111
   band 111
   combination 111
   crosscut 111
   hollow-round 111

power tools 111
  rip 111
  stationary 111

Saws 8, 11, 24, 36, 40
  back saw 11, 18
  band saw 40, 44
  bench saw 40
  chain saw 36
  circular saw 30
  compass 24
  coping saw 11
  crosscut saw 8
  drill 40
  electric 33
  hack 24
  jig 34, 40
  keyhole 57
  radial arm 40, 42
  rip 8
  reciprocating 36
  saber 34, 57
  table 40

Saws, sharpening 108, 109

Screwdrivers 12
  cabinet 12
  jewelers 12
  offset 12
  Phillips 12
  screw holding 12
  slim 12
  standard 12
  stubby 12

Screws 94, 95
  lag 95
  Phillips 94, 95
  self-tapping 95

Scroll saw 47

Shaper 40, 48
  decorative beads 48
  fancy cuts 48

Sheathing 81

Shellac 116, 117, 121
  orange 117
  quantities 117
  white 117

Shelves 112
  angle irons 112
  brackets 112
  dowels 112
  heavy-duty 113
  short shelves 113
  steel rods 112

Skylights 83

Soffit 89

Splices 70

Square cuts 38

Squares 14
  combination 14
  miter 14
  roofing 14
  try 14

Staining 115, 116
  Japan colors 116
  water-soluble 115

Storage 65

Subflooring 78, 80
  common 80
  plywood 80
  shiplap 80
  tongue and groove 80

Surform tool 13, 24

Switch plate 56

Tongue and groove 81

Toning 116, 117

Tools 108
  sharpening 108, 109

Trim 123

Undercoat 118, 121

Utility knife 22

Varnish 116, 117
  polyurethane 117
  "spar" 117

Ventilation 58

Vises 20, 65
  machinist's 21
  woodworker's 21

Wallboard 60, 81

Wall scraper 22

Waterproofing 58

Whetstone 24

Wing divider 22

Wood 52-54
  board feet 52
  common 52
  estimating 52
  hardwood 53
  plywood 54
  select 52
  softwood 53

Wood paneling 56
  installation 56

Workbench 26, 64, 65, 68
  parts 69

Workshop 64
  floor plan 67

Wrenches 22
  box, open end 22
  open end 22
  pipe (Stillson) 22